SCIENCE ON A SHOESTRING

Herb Strongin

 ADDISON-WESLEY PUBLISHING COMPANY

Menlo Park, California ● Reading, Massachusetts
London ● Amsterdam ● Don Mills, Ontario ● Sydney

This book is published by the Addison-Wesley Innovative Division.

Published simultaneously in Canada.

ISBN 0-201-07329-3
LMNO-SE-98765

Contents

THEME 3

What Is Science on a Shoestring?

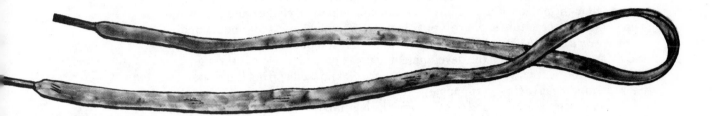

The *Science on a Shoestring* program is designed for the K-7 teacher with little or no science background, few supplies and materials, an extremely limited budget, and an ordinary classroom. It enables teachers to bring an exciting and significant hands-on program to their classroom or school.

Several years ago the San Francisco Unified School District attempted to find out what was really happening in science K-12. It was discovered that a few teachers had favorite science units they taught well each year but, generally, although science was required, it was possible for a youngster to go from kindergarten through sixth grades without any consistent science program at all.

Why wasn't science being taught? The teachers shared these reasons:
- The textbook was too difficult to use.
- The teachers didn't have the materials for hands-on science.
- There just wasn't enough time in the day to teach science.
- Teachers felt very uncomfortable about teaching something that they knew little about.

In order to overcome all of these problems, we in the district first had to decide what is important to teach in elementary school science. There are three important components:

1. knowledge 2. concept development 3. scientific process.

Knowledge is the facts, the vocabulary, and the key ideas that are necessary to develop and communicate scientific ideas.

The *concepts of science* are the broad fundamental understandings that must be developed over a long period of time before they can be internalized and really understood.

The *processes of science* are the techniques that are used to develop and test scientific knowledge and concepts.

The processes include observing, recording, hypothesizing, experimenting, evaluating, predicting, and retesting.

Of the three components of an elementary science program, the knowledge component is the *least* effectively and efficiently carried on by the teacher. Library books and textbooks are filled with knowledge that youngsters can obtain if they are so motivated. Television provides a wonderful opportunity for the development of knowledge in home or at school.

There is no one on TV, however, who can do a better job than the teacher in helping the child to develop the processes and concepts or science. With just two 20-cent magnets and thought-provoking questions a teacher can help to place the child center stage in one of the most exciting endeavors possible, the individual quest for understanding.

Science on a Shoestring—the program written as a result of the San Francisco study—therefore focuses on the development of concepts and processes through hands-on activities that will motivate children to seek knowledge from the resources that are available in their school, home, or community.

To accomplish these goals and to overcome the legitimate objections that teachers have for teaching science, the program has been designed as follows.

First, the *program is student centered.* It requires no textbook but may be used in conjunction with any textbook.

Second, the *program uses inexpensive, readily available materials.* The low cost helps put the tools of science in the hands of the youngsters. The common nature of the material permits the student to develop concepts without having them obscured by the equipment.

Finally, the book was developed as a tool that would enable the teacher to easily initiate the science activity.

How To Get Started

If you are the only teacher using the Science on a Shoestring (SOS) program:

Select the materials for the activities you intend to do. Just about everything you will need is available at local stores. The few exceptions, such as test tubes and litmus paper, can be found in hobby shops or science supply stores. Learning Spectrum carries all SOS materials and will supply complete, reasonably priced SOS kits and replacement materials (see the order form following page 28).

If many teachers at one school or district are using the SOS program:

It would be best to plan to use the SOS activities so that students won't repeat them each year, as follows:

● Review your existing science program.

● Determine the grade level at which each hands-on SOS activity best fits.

● Obtain materials for each activity and place them in a shoebox or similar container.

● Distribute appropriate boxes to each teacher responsible for implementing specific activities. The boxes should be left in the classroom with the teacher. This encourages flexibility and spontaneity. The few expensive items, such as SOS microscopes (see insert following page 28) and plastic shoes boxes in class quantity, may be stored in a central location for use as needed. Replacement items should be purchased in advance and stored centrally.

We've found that even the teacher who is reluctant to teach science will be able to offer a significant and exciting hands-on program when supplied with materials, given this easy-to-use guide, and asked to teach only a few specific SOS activities.

Using the Investigations

The investigations are grouped under three conceptual themes: 1. scientific methods; 2. change; 3. energy, fields, and forces.

Each conceptual theme represents a major area of science and will be discussed at further length as it is introduced in the program. The boundaries of these areas are not fixed, and eventually the youngsters will understand that the major themes overlap and are really part of a unity that exists in science and in the environment.

It is suggested that the youngster move toward this "big picture" in steps that they and the teacher can handle. These steps are taken during the student investigations. Most investigations may be introduced without regard to sequence. A few, however, are sequential. This is indicated under the heading *Note to Teacher*. The objectives of each investigation are listed under the heading *Specific Concepts/Skills*. Student progress is measured in a variety of ways suggested under the heading *Evaluation*. There are also suggested ways for the teacher to determine his or her own progress.

A typical investigation begins with the teacher introducing the problem to the class under the heading *The Activity*. The "script" spells out how the teacher may do this. After the problem is understood, the materials are distributed, and the students begin their investigations. Rarely do two classes or even two students react in the same way. Consequently, the script can only generalize what is likely to occur thereafter.

Youngsters who have not often been given the opportunity to work things out may just putter with the materials while waiting for you to give "the" answer. This style of problem solving has been learned through years of TV watching. A typical TV program begins with a problem, continues while the actors attempt to resolve it, and concludes with a solution, all within the preannounced time module. Commercials do this too, and usually within a minute.

Problem: Dirty clothes. Investigation: Try Flab soap. Conclusion: Clean clothes with Flab soap.

In the fields of science and environmental education TV does certain things well. It produces outstanding "show and tell" programs that move the youngster vicariously from the miniscule to the infinite, from complex problem to brilliant solution, without taxing the child physically or mentally.

It is not surprising, then, that many youngsters will be annoyed when science period is over and you haven't given the answer.

"What kind of program is this? I sat here for 30 minutes, so why don't you give me the answer?"

The teacher must resist becoming the answer person. He or she must bring the youngsters center stage and help them to solve problems by developing their processes of investigation, including questioning, observing, recording, interpreting, hypothesizing, testing, sharing, discussing, and being patient. For, after much careful study and questioning, some answers may just not be found.

Elementary students are not famous for their patience and long attention span.

Consequently, most investigations are designed to be completed within 30-60 minutes. The suggested time needed is listed under the heading *Time*. Additional activities are suggested for youngsters capable of extended investigations.

The few investigations under the category *More Activities* require some expensive or hard-to-find materials such as microscopes, vacuum pump, fishing rod, etc. These few items may be obtained on loan from a neighboring junior or senior high school or may be purchased for use by your entire school.

Obviously, skills and concepts cannot be learned well during one 30-minute module. Reinforcement is necessary. This is achieved by focusing attention on similar concepts and skills during subsequent investigations, and by permitting youngsters to perform investigations suggested under *Investigations at Home*. Investigations at home are particulary useful as they not only allow youngsters to apply what they have learned, but they also generate a positive interaction between school activities and the youngsters' home life.

Scientific Methods

All the investigations in this book require the use of scientific methods and are intended to help develop scientific attitudes. The specific investigations placed into Theme 1, Scientific Methods, are particularly structured so as to stress use of skills and development of attitudes that have proved useful in problem solving.

The investigations confront the youngsters with problems whose solutions require scientific methods such as observation, data gathering, communication with peers, manipulation of equipment, hypothesizing, prediction, and experimentation.

Scientific methods should not be confined to investigating science problems. Rather, they should be applied to the everyday decisions a youngster will make. This will reinforce and give meaning to what the students have learned in class.

You as the teacher can pick out those common problems that lend themselves to scientific solutions. Then, you can guide youngsters to applying scientific methods in resolving the problems.

For example, you can have the class evaluate common TV commercials that ''prove'' their product best by using pseudoscientific experiments, generalizations, false authorities, and catchy and meaningless slogans.

A trip to the supermarket may enable youngsters to compare similar products as to container weight, volume, nutritional content, and price.

A teacher can encourage youngsters to make decisions affecting classroom management by applying scientific methods. First determine the problem, then use observation, data gathering, and communication to determine the available options. Hypothesize a solution, and then evaluate results, modifying the solution where necessary.

Youngsters will gradually develop an attitude toward problem solving that will enable them to decide when to apply scientific methods. After all, some problems are better solved with the heart than with the mini-calculator. But far too often, decisions are made without use of scientific methods only because the decision maker lacks experience and skill in using them.

MATERIALS
Per class
1 bottle red food coloring
1 bottle green food coloring
3 eye droppers
few drops water softener
Per team
3 waxed containers or 3 pieces
  (13 × 13 cm or 6″ × 6″) wax
  paper
*1 multipurpose plastic
  container cover (MPC cover)

VOCABULARY
(optional) surface tension

Investigating Red Drop - Green Drop

Specific Concepts/Skills

1. This lesson is intended to help youngsters develop observation skills.
2. (optional) The surface tension of water is reduced by the addition of a water softener.

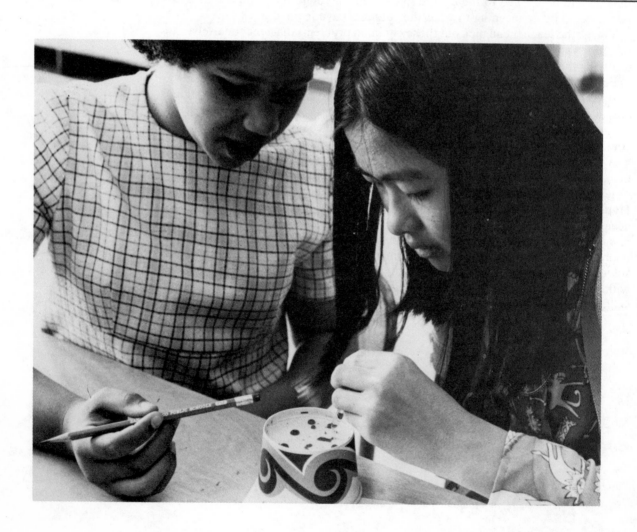

Advanced Preparation

This lesson is easy to prepare, great fun, and good science. Don't pass it by!

Prepare one bottle of Red Drop and one of Green Drop in advance. To make Red Drop, add a few drops of red food coloring to a small container of water. Be sure the water is bright red. For Green Drop, add a few drops of green food coloring to water, plus one or two drops of water softener. (Photo-Flo, available from most camera stores, works particularly well. One capful makes a half gallon of very soft water. However, any softener or detergent will do.)

Test the two liquids before class. Red Drop should form a rounded drop on a waxed surface. Green Drop is blah—it should just lie flat on a waxed surface. If it doesn't just add a drop or two more water softener.

The Activity

1. Pass out the materials. Best results are obtained using the back side of a waxed container such as a potato salad cup. However, waxed paper or a cut-down milk carton will do.
2. If you use waxed paper, have the youngsters place one piece on the MPC cover*. The MPC cover will support the paper and allow the youngsters to move the drops around freely.
3. Tell the class you are going to place a few drops of red liquid on the waxed paper. They are to discover all they can about Red Drop and list its properties on a worksheet.
4. Place about half a dropper of Red Drop on each piece of waxed paper. Allow sufficient time for discovery. Some typical descriptions are:

   It's round but when it moves it stretches like a tear drop.
   Big drops swallow the little ones.
   It follows my eraser.
   It has no smell.

5. Ask the youngsters to carefully move the waxed paper containing the Red Drop from the MPC cover to their desk. Have them place a second piece of waxed paper on the MPC cover.
6. Tell the class that you will now place a few drops (*be sure to use a clean eye dropper*) of a green liquid on their waxed paper. They are to discover all they can about Green Drop. Again, allow sufficient time for discovery. Some typical descriptions are:

*This multipurpose plastic container and its cover (MPC and MPC cover) will be used frequently in SCIENCE ON A SHOESTRING. A clear plastic shoe box is best.

It spreads out.
You can't pull it around.
It's no fun.
When two drops touch, they join.
It's like water.

7. Ask the youngsters to transfer a bit of Green Drop to the Red Drop. Ask them to list the properties of the merged drop. (The combined drop behaves as Green Drop.)

8. Tell the class you are going to place a few drops of plain water on their last piece of waxed paper. But first, have them predict which drop (Red or Green) behaves most like water. (Most will predict Green Drop.)

9. Place about half a dropper of water on the third piece of waxed paper. Again, allow the students time to observe and compare. (Water and Red Drop behave identically.)

10. Allow the youngsters to continue investigations. Let them try both Red and Green Drop on different kinds of surfaces. Conclude with a group discussion of the properties of each substance. You may want to explain how Red and Green Drop were made.

11. (optional) The water softener breaks the water's surface tension and it flattens out. The pure water and Red Drop have surface tension and bunch up on waxed paper.

12. Conclude by asking such questions as:

    a. How does water on a waxed surface differ from water on an unwaxed surface? (Bunches up more on a waxed surface.)
    b. How do you think waxed paper helps keep food fresh? (Holds in water.)
    c. (optional) How does a water softener (or detergent) help keep water spots off dishes? (Softener causes water to run and not bunch up. Water spots are deposits of minerals left after water drops evaporate.)

Investigations at Home

See More Activities for Theme 1, Lesson 1M (page 5)

Evaluation

1. Did the students observe significant differences between Red Drop and Green Drop?

2. Did the students want to continue to "play" with Red Drop? Did you let them? (You shouldn't; just move on to the next investigation.)

3. (optional) Were most students able to explain the fall of Red Drop (when Green Drop was added) using the term surface tension?

MATERIALS
2 clear plastic cups
2 needles
detergent or Photo-Flo

Floating a Needle that No One Else Can

What to Do

In this investigation you will float a needle on water, but no one else in the class will be able to do it.

Announce that you have the ability, intelligence, and necessary skill to float the needle in the glass of water. If anyone in the class can get the needle to float, you will award that person a great and magnificent prize. Carefully lower the needle onto the surface of the water. (Practice beforehand to be sure you can do it. One way to get it to float is to place tissue paper on the water, then place the needle on the paper, and encourage the paper to sink. The needle will float.) Once the class has shown proper WOW, remove the needle and ask for volunteers. No one will be able to get it to float.

The trick is that before you begin, you place a bit of detergent under the nail of your middle finger. Float the pin using thumb and index finger. When you remove the pin use the middle finger and thumb. The detergent will quickly dissolve in the water and break the surface tension. Since the water's surface tension supports the needle, no student will be able to get the needle to float.

If students ask you to do it again, use fresh water and a new needle.

Tell them to observe what you do very carefully. After the trick is discovered, ask them to practice floating a needle at home, and see it they can fool their friends or parents.

You may wish to review Red Drop—Green Drop (and surface tension), using the trick as a motivator.

Incidentally, a favorite prize is a "solid gold" tie clip for boys (a brass paper clip) or an organic perfume factory with metallic fastener for girls (a flower with a straight pin).

For further interest you might ask the following:

1. Has any one ever seen an insect walk on water? How do they do it? (Water appears to bend under the legs; the water's surface tension supports them.)
2. What would happen if a water softener were added to the insects' water? (It would sink.)

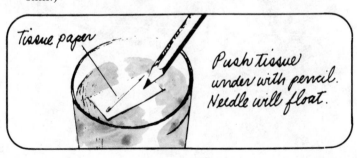
Tissue paper

Push tissue under with pencil. Needle will float.

Investigating Your Five Senses Using Sense Boxes

Specific Concepts/Skills

1. We have five senses which help us investigate our environment: touch, sight, smell, taste, and hearing.

MATERIALS

Per team

1 mystery sense box (see page 9 for instructions on building)

masking tape

1 straw per student

1 answer sheet (see page 8 for model)

VOCABULARY

The sense of: touch, taste, sight, hearing, smell; environment

Note to Teacher

Although this lesson requires more preparation than most others in the book, I think you will find it worth using. The instructions for building sense boxes and a sample sense box answer sheet are on pages 9-10.

The Activity

1. Say: "Each of us can use our five senses to investigate our environment." If necessary, review the concepts of environment and the senses. (Environment is everything around us. Senses are parts of our body especially adapted to receive information from our environment.)
2. Ask: "Who knows what our five senses are?" (With younger children, drawings illustrating the main locations of the five senses are helpful.)
3. Discuss the five senses and their locations. (Hearing, tasting, seeing, and smelling are located in specific places; touching, or feeling, senses are located all over the body.)
4. Say: "We are going to use each of the senses to investigate the mystery sense boxes."
5. Pass out the mystery sense boxes, giving at least one to each team of three.
6. Say: "Please don't touch the boxes until I explain what you are to do. Your job is to find out what is in the box, using *only* the sense written (or drawn) on the box." Illustrate what you mean using one box.
7. Pass out an answer sheet to each child.
8. Say: "Write down (or draw) on your answer paper what you think is in the box. When I say time is up, pass the mystery sense box on your table to _____. Some of the boxes call for the sense of hearing. So, you must be very quiet. Any questions? Let's begin."
9. When all teams have investigated each box, it is time to review. Allow the children to keep their answer sheets and discuss each box. Don't open them. Giving the answer is too final and tends to end discussion.
10. Review the lesson.

Investigations at Home

Ask the children to make their own mystery sense boxes and have them bring them to school. You may use them to extend the lesson an extra day. (You may also get some good ideas for next year.)

Evaluation

1. Did the children enjoy the lesson?
2. Can every student name the five senses and their locations?
3. Do students use the word *environment* comfortably?
4. Did they cooperate; was the classroom reasonably quiet?
5. Was the building of mystery sense boxes worth the effort?
6. Did the youngsters want to continue the lesson the next day? Did you let them?

ANSWER SHEET FOR INVESTIGATION 1-3
Sense of smell What was in the box? *box 1*
box 2
Sense of sight *box 3*
box 4
Sense of taste *(box 5)* *container A*
container B
container C

Instructions for Building Mystery Sense Boxes

What follows will require some effort, for which you will be amply rewarded by the enthusiasm and delight of your class. Once made, sense boxes are reusable, so perhaps other teachers will help share the work and the rewards.

Shoe boxes are best but cereal or detergent boxes will do. The contents of the sense boxes described below have already been classroom tested, but don't use anything you feel uncomfortable with. Substitute with anything you want.

Number the boxes consecutively. This will help identify a particular box for discussion and placement on the answer sheets.

Sense of Smell

Tape an onion and a small sponge sprinkled with toilet water or aftershave lotion to the bottom of a box. Poke three or four holes in the cover. Tape the cover shut. Tape the following instructions to the cover: "Use only your sense of smell to find out what is in this box." For younger children use oral instructions and draw a symbol (a nose) representing the sense to be used.

Prepare other boxes as above and include such items as library paste, clay, flowers, or various foods.

Sense of Sight

Place a potpourri of common items in a box. For example, crushed cornflakes, a variety of common seeds and nuts (apple, pear, plum, etc.), carrot tops, celery tops, chalk, tobacco, etc.

Cover the box with a clear plastic wrap and tape the wrap securely.

Instructions: "Use only your sense of sight to identify as many things as you can in this box." For younger children, give oral instructions and tape a drawing of an eye to the lid.

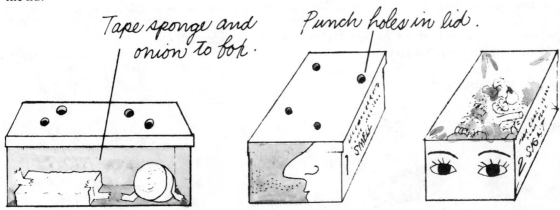

Tape sponge and onion to box.

Punch holes in lid.

Sense of Touch

Place several items in each box. Each item should be placed in a plastic or cloth bag and secured. Use, for example a marshmallow, Jell-o, a cotton ball, raisins, dry cereal, and marbles. Cut a small opening into the side of the box and staple cloth to cover the opening. Tape the cover on.

Instructions: "Use only your sense of touch to find what is in the box. Do not take anything from the box."

Sense of Hearing

Tape a ticking alarm clock into a box (for added interest, set it to go off during the investigation). A bell, rattle, or living animal (such as a mealworm, or jumping bean) can be used. For something exotic, tape an egg timer to the bottom of the box. Tape the box closed.

Instructions: "Turn the box over and use your sense of hearing." Tape the instructions on both sides.

Sense of Taste

Prepare weak solutions of salt and water, sugar and water, lemon and water, coffee and water, etc. Place the solutions in washed milk cartons or other suitable container. Punch a hole in the top of each carton large enough to admit a straw. Number each carton and put them all in an open box.

Each youngster trying this investigation should be given a clean straw. (Some teachers prefer to use different solid foods such as raisins, nuts, lemon peel, salt, or sugar.) Have the youngsters taste one item and describe the flavor.

Using Any or All Senses

A very interesting sense box can be made using a balloon partially filled with water and tied shut, a peanut or a small rock, and a piece of onion. Place all three in a box and tape it shut.

Instructions: "Use all of your senses to find what is in the box, but don't open it." For the younger children, draw all the senses.

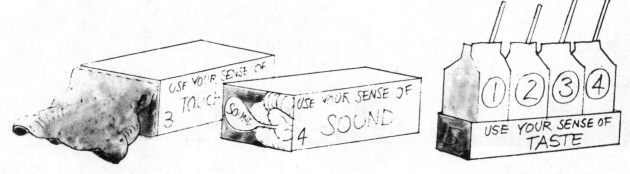

MATERIALS
Per class
eye chart (optional)

VOCABULARY
three dimensions

Testing Your Senses

Specific Concepts/Skills:

1. Both eyes are necessary for seeing in three dimensions.
2. Both ears are needed for determining where a sound is coming from.
3. The nerves of touch are not distributed evenly throughout the body.

Note to Teacher

This lesson continues the investigation of the senses. The suggested activities are grouped by the sense to be discussed.

The Activity

Review the previous lesson.

Sense of Sight

1. Ask each student to face another person. Have each partner cover one eye. Then tell them to extend the index finger of the free hand toward each other so that their fingers meet in mid air. (Chances are they will miss.)
2. Let the class have several tries. (No need to say anything; they'll keep trying until they hit.)
3. Ask them to open both eyes and try again. (This time they'll have no trouble.)
4. Discuss why both eyes are needed. (With only one eye, you see only in two dimensions. Both eyes are needed for three-dimensional sight.) In two dimensions you can only see left and right and up and down. You cannot see depth. A picture on a TV screen is in two dimensions. However, the illusion of 3-D is made when approaching objects get larger and receding objects get smaller.)
5. Hold up an eye chart and ask: "What is this used for?" (To determine how well each eye sees at a certain distance.)
6. Say: "If you want, I will check your eyesight." (Follow instruction on back of the chart or have the school nurse do the testing.)

Sense of Touch

7. Ask for a volunteer who will turn his or her back to the class. Then press five fingers against the child's back. Ask how many fingers are felt. (If your fingers are close together the youngster may only feel two or three fingers.) Repeat using different numbers of fingers. Have the volunteer place his or her hand behind his or her back palm up and press your fingers on the palm. (The youngster should easily determine the number of fingers you are using.)
8. Have the class pair off and try the investigation.
9. Discuss the results. (The nerves that sense touch are closer together in some parts of the body. For example, just a speck that can't be felt by the hand can be excruciating in the eye.)

Sense of Hearing

10. Say: "You are going to test your sense of hearing. Sit very quietly, without moving or speaking. Listen for any sounds. After five minutes (more time if you have a

headache), I will ask you to write down (or describe) all you heard."

11. Say: "We have learned that two eyes are necessary for seeing in three dimensions. Do you think two ears are necessary for hearing in 3-D?" Allow discussion and review what three dimensions are: up — down, left — right, and forward — backward.

12. Ask for a volunteer. Have the person come to the front of the room and turn her or his back to the class. Then ask the volunteer to cover one ear tightly.

13. Tell the class that when you point to a student, you want that student to tap on his or her desk or snap his or her fingers. The volunteer is to guess where the sound is coming from.

14. Repeat but have the volunteer listen with both ears.

15. Discuss the results. (Both ears are needed to pinpoint sound direction.)

Senses of Smell and Taste

16. There are many ways to investigate the senses of smell and taste. For example, plotting the taste buds on the tongue (sweet, sour, salt, bitter), or saturating the olfactory nerves with one aroma and preventing the sense of other odors. But for younger children a story and the discussion that follows is recommended to wind up (for now) the study of senses.

STORY: Many years ago there was a city in the Middle East where many poor people lived. One especially poor beggar almost never had anything to eat except bread. However, he had discovered that if he sat just outside of a delicatessen, he could smell all the wonderful smells of cooking meat and, when he ate his bread, it tasted as though he were eating the delicious meat. He would come and sit outside the delicatessen almost every lunch and dinnertime.

Well, it didn't take long for the delicatessen owner to realize what was happening. He soon had the beggar arrested and charged with stealing the smells of his delicatessen.

Within a day the poor beggar was brought before the king for trial. He was very nervous because the punishment for stealing was the loss of both hands!

The king asked the delicatessen owner to explain the charge, which he did. The king then asked the beggar, "Did you smell the aroma of cooking meat to make your bread taste better?" The poor beggar said he did. "Then," said the king, "you are guilty and you must pay the delicatessen owner 20 gold coins or lose your hands." The delicatessen owner was pleased but the beggar had no gold coins and was very sad. The king said, "I shall lend the gold coins to the beggar," and called the delicatessen owner to his throne. Then he took out 20 gold coins and one by one he dropped them on the floor in front of the store owner. "Did you hear all 20 coins hit the floor?" said the king. "Yes," said the smiling store owner. "Good," said the king, "then you are repaid by the *sounds* of the coins for the stolen *smell* of the meat." The smile left the face of the store owner as he returned the coins to the king.

Discussion

a. Did the beggar really take anything from the delicatessen owner? (Yes, the smell of meat, but is that stealing?)
b. Do you think the king was fair?
c. Do you believe that if you ate bread, while smelling meat, the bread would taste like the meat? (Try it.)
d. If you held your nose, so you couldn't smell what you were eating, could you taste it? (Try it; you can only taste sweet, sour, bitter, and salty—and therefore most of the subtle flavor is lost because you can't smell the food. As you know, when you have a cold nothing tastes very good.)

Investigations at Home

Encourage the students to repeat the investigations at home, using their family as subjects. Have them report results in class.

Evaluation

1. Do most youngsters realize the need for both eyes and ears for three-dimensional sensing?
2. Did the youngsters want to try the tests on their partners in class and friends out of class?

MATERIALS
Per class
1 balloon
1 clear plastic cup
food coloring
Per student
3 small plastic bags
3 twist bands

VOCABULARY
solid, liquid, gas,
invisible,
visible

Investigating and Classifying Solids, Liquids, and Gases

Specific Concepts/Skills

1. Most things are found as a solid, a liquid, or a gas.
2. Substances may be classified as solid, liquid, or gas according to their properties.

Note to Teacher

In this lesson youngsters observe the properties of matter and classify different substances as solids, liquids, or gases. Not all things will neatly fit these categories and you may want to create a "doesn't fit" category. In the next lesson, students discover a substance that has the properties of all three: solid, liquid, and gas. Their reaction to this anomaly should provide you with an opportunity to emphasize creative thinking.

The Activity

1. Say: "We are going to study solids, liquids, and gases."
2. Use three plastic bags. Fill one with any handy solid. Tell the children that _____ is a *solid*.
3. Ask: "Can you see this solid?" Introduce the word *visible*. "Does the solid change shape easily?" (No, it holds its shape.)
4. Pour the solid on the table.
   Ask: "Does its shape change easily? (No, not easily. A rock may be broken and its shape change but the rock does not change shape by itself.)
   Ask: "Can I move a pencil easily through it?" (No.)
6. Summarize the properties of a solid.
7. Fill a plastic bag with water. Tell the children the water is a *liquid*.
8. Ask: "Can you see this liquid?" (Pure water is *invisible;* it will, however, bend light [refract] and reflect color.)
9. Add a few drops of food coloring to the water in the bag so the children can follow its movement. Use a twist band to close the baggie.
10. Move the water around by tipping and moving the bag.
    Ask: "Does the liquid change shape easily?" (Yes.)
11. Pour the water from the bag to another container. Have the children notice that, as the liquid is poured, it changes its shape.
12. Ask: "Can I move a pencil easily through the liquid?" (Yes.)
13. Summarize the properties of a liquid.
14. Ask: "What is right in front of your nose?" (Air.)
15. Children may say "nothing." If so, have them wave their hands in front of their face; they will feel the air.
16. Blow in a plastic bag and tie it shut.
17. Ask: "What is in the bag?" (Air.)
    Tell students that air is a *gas*.
18. Ask: "Can you see air?" (Pure air is invisible, but polluted air is not.)
19. Ask: "Does the air change shape?" (Yes.)
    Open the bag. Most of the air in the bag joins the air in the room and therefore changes shape.

20. Ask: "Can I pass a pencil through the air?" (Yes.)
21. Review the similarities and differences of solids, liquids, and gases.

Solid
1. Doesn't change shape easily.
2. Another solid can't be passed through it easily.
3. Usually visible.

Liquid
1. Changes shape easily.
2. A solid passes through it easily.
3. May be visible or invisible.

Gas
1. Changes shape easily.
2. A solid passes through it easily.
3. Usually invisible.

22. Have the youngsters prepare a list of solids, liquids, and gases found in the classroom.

Investigations at Home

Pass out three plastic bags and three twist bands to each youngster. Tell them to find (at home) a solid, a liquid, and a gas, and to place one in each bag. Next day, allow the children to show their bags to the class. Have them tell whether the bags contain a solid, a liquid, and a gas, and why. Review again the properties of solids, liquids, and gases.

Evaluation

1. Were the children able to identify solids, liquids, and gases in the room and at home?
2. Do the words *solid, liquid,* and *gas* appear in students' vocabulary?

MATERIALS
Per student
1 small piece of Silly Putty

Investigating a Solid / Liquid

VOCABULARY
properties

Specific Concepts/Skills

1. Some substances have the properties of both solids
   and liquids.

Note to Teacher

In this lesson, youngsters review the properties of solids, liquids, and gases. They will
find that not everything is easily classified.

The Activity

1. Say: "Today we are going to study the properties of an interesting material. What do you think the word *properties* means?" (Properties are the important characteristics of something. Allow the youngsters to explain in their own way.)
2. Review the properties of a solid, a liquid, and a gas (Lesson 1-4).
3. Give each child a piece of Silly Putty and retain a large piece for yourself.
4. Say: "Find out as many properties of this material as you can. Remember, your job is *not* to guess what it is, but to find and describe its properties." Youngsters may discover:

> If you stick it on the wall, there won't be a stain.
> It's like clay but different.
> Water can't get into it.
> It picks up what your skin looks like.
> It picks up letters from the newspaper and makes them backwards.
> It floats. It doesn't float. (Depends on shape.)
> It stretches like a rubber band, but it doesn't go back.
> When you drop it, it bounces no matter what its shape is.
> When you heat it, it turns brown and sticky.
> When you rub it, it gets shiny.
> It smells good.

5. Place something on your Silly Putty, such as a coin, a pencil, or a clothespin.
Ask: "What is happening to this Silly Putty?" (It changes shape, it flows within a minute or two.)
6. Place the words *solid* and *liquid* on the board. Ask the class to recall the properties of each and write them under the words. Then, ask for volunteers to describe one property of Silly Putty that they discovered.
Ask: "Do you think that is a property of a solid or liquid?" (Encourage discussion but stop fist fights!) Continue listing all properties discovered by students.
6. Conclusion: This material has the properties of both a solid and a liquid.

Evaluation

1. Did the youngsters use the word *properties* correctly?
2. Did most youngsters describe significant properties?
3. Did most youngsters come to the conclusion that the material had properties of both a solid and a liquid?
4. Did they want to keep the Silly Putty? Did you let them?
5. Did you know that when Silly Putty gets into woven things, you can't get it out?

Making Your Own Candle

Specific Concepts/Skills

1. A candle is made of a wick and wax.
2. A burning candle is a source of heat.
3. Wax changes to a liquid when heated and back to a solid when cooled.

Note to Teacher

This lesson is presented here for two reasons: 1. The recommended source of heat in this book is the candle. It therefore seems sensible that the youngsters become familiar with the candle and how it is made as soon as possible. 2. While making and then burning a candle youngsters can review the properties of solids, liquids, and gases.

Advanced Preparation

1. Prepare seatwork for children not dipping candles. (In smaller classes, all may dip.)
2. Prepare cord for wicks (two days before project) or purchase commercial wick, or use just plain string. It is not as good, but it works.

   a. Cut cotton cord in about 40-cm (16-in.) lengths for double-strand wick or 25-cm (10-in.) lengths for single-strand wick.
   b. Soak cord overnight in the following mixture: 2 tablespoons borax, 1 tablespoon salt, 1 cup water (mixture should cover cord).
   c. The next morning hang the cord so it can straighten out as it drips dry.
3. Prepare wicks for dipping (one day before project):

   a. Using masking tape, mark each pencil with child's name.
   b. Loop one end of wick over middle of pencil and tie securely.
   c. Fasten paper clip to other end of wick. (This weight keeps the wick straight during the first candle dippings.)
4. Prepare wax for dipping (before class):

   a. Cut or shave wax into 46-oz juice can. Do same for some colored crayons.
   b. Place juice can in a pan of water and heat slowly until wax melts. (This may best be done in teachers' lunchroom just before class so there is no hot plate to worry about around the children.)
5. Arrange an area ahead of time where two ends of pencil may rest on two chairs or tables, etc., and candles hang straight down to harden.

The Activity

1. Before beginning this project with class, caution children about *safety first. Hot wax and hot water can burn. No one runs. Nobody pushes. Only children dipping candles can be out of their seats, and they must take turns.*
2. Explain that one half of the boys and girls will stay in their seats and work on something while the other half of the class dips candles. Tomorrow the other boys and girls will have their turn dipping candles.
3. Pass out and explain seatwork to those children not dipping candles.

4. Have children dip wick into liquid wax and remove. They then should hold a piece of cardboard under the candle wick so no wax drips on the floor.
5. After a number of dippings, candle will stay straight and paper clip should be clipped off bottom of candle. It is best to do this before finishing the candle.
6. A candle one-half centimeter (one-fourth inch) thick is sufficiently large to use as a heat source.
7. Review the children's candle-making process and discuss how candles were made many years ago.
8. Review the changes of state involved in candlemaking.
9. Candles may be used in class or sent home.

Evaluation

1. Did the youngsters want to go on and on making an even thicker candle?
2. Do they think you are a wonderful teacher? (You are.)
3. Every once in a while did you hear the words *solid* and *liquid* (even if you had to say them)?

MATERIALS
Per team
1 multipurpose plastic
　container (MPC)
1 clear plastic cup per student
masking tape
newspaper or paper towels
sponge for spills

VOCABULARY
air (the concept), laboratory,
predict, hypothesis

Investigating Air and an Empty Cup

Specific Concepts/Skills

1. Scientists must work in a proper environment
2. Air is all around us.
3. Air takes up space.
4. Air and water cannot be in the same place at the same time.

Note to Teacher

All investigations in this lesson require one MPC filled with water for each team. Many classrooms have water and youngsters can easily fill the MPC and carefully return it to their desk or other work area. However, if your room lacks water, a plastic pail (costing about $1.50) can be filled with water from the nearest source and used in your room. Just have the youngsters dip their MPC into the pail and fill. Have a sponge available to wipe off excess water from the outside of the MPC.

The Activity

1. Explain to students that when they carry out a science investigation, their room becomes the science laboratory and their desk the work area. It is important that the work area be kept clean, and that the laboratory remain quiet so that the ''scientists'' can think and work. Scientists do talk to each other, but it must be done quietly so as not to disturb others.

2. Say: ''Please prepare your work area by removing everything from your desk.'' (Some teachers have students place newspaper or paper towels on the desk to absorb spilled water.)

3. Have students (''scientists'') fill their MPC about 2 centimeters (an inch) from the top and return it to their work area. (Young children may have trouble carrying an almost full MPC. You can have them fill it about halfway and then you top it off with water carried in a portable container.) Distribute one clear plastic cup to each child.

4. Ask: ''What do you think is inside your cup?'' (Nothing—light—air.) ''Today we are going to try to find out what is inside the cup, if anything.''

5. Have each youngster feel the inside of the cup with *dry* hands. Ask: ''What do you feel?'' (Nothing.) ''Is it wet or dry inside?'' (Dry.)

6. Tell them you want them to do something after they see you do it. Take a small piece of paper towel or newspaper and crinkle it up. Place it in the bottom half of the plastic cup. Turn the cup over and shake it a bit. If the paper falls out, repeat; but this time, tape it in place with a bit of masking tape.

7. Distribute the paper towels or newspaper and a small piece of tape if necessary. Caution the children not to wet the cup or paper. Have them turn the cup over and shake to be sure the paper will stay in.

8. Tell the children to observe what you do. Hold your cup upside down over the MPC.
   Ask: ''If I lower the cup straight down, what do you think will happen?'' (Allow for discussion. You may wish to tell them that their guess is called a hypothesis.) Say: ''Lower your cup as I do and hold it at the bottom of your MPC. Has water entered

the cup?'' (It's hard to tell.) Say: ''Carefully raise your cup straight up and out of the water. Wipe extra water from the outside. Look inside. Did water enter the cup?'' (No.) ''How do you know?'' (The paper is dry.) Say: ''Repeat the investigation to see if you get the same results.''

9. Ask: ''Why do you think the water did not enter the cup?'' (Allow children to give their own hypotheses.)
10. Say: ''Let us try to find out. Place the cup straight down as you did before. Now carefully tip the mouth of the cup up just a little.''
11. Ask: ''What do you observe coming out of the cup?'' (Bubbles.) ''What do you think the bubbles are?'' (Water bubbles, air bubbles, I don't know.)
12. Say: ''Continue to tip your cup and observe what happens.'' (More bubbles leave and the paper gets wet.)
13. Ask: ''In order for the water to enter the cup, did something have to leave the cup first?'' (Yes, bubbles.) ''What are these bubbles made of?'' (Air.)
14. At this time many younger students may not really understand that air kept the water out of the cup. Lesson 1-8 continues the development of the concepts, so don't be concerned if there is more enthusiasm than there is understanding. By the end of Unit 1, most youngsters should have a feel for the concepts.
15. Review. If time permits you may want to go on to Lesson 1-8 as all needed materials are at the students' desks and 1-8 reinforces concepts learned in 1-7.

Investigations at Home

Ask the youngsters to repeat the investigations at home using a glass and their sink or a deep pot. Ask them to see if their parents or friends can predict what will happen. Allow for discussion next day.

Evaluation

1. Were the youngsters able to follow directions?
2. Did they observe carefully?
3. Were they able to verbalize their hypotheses and observations?
4. Did they enjoy the lesson? Do they look forward to science period? Do you? (It is a lot of work, but it can be worth it.)

MATERIALS

Per class
food coloring
Per team
1 MPC
2 clear plastic cups

VOCABULARY

transfer, dissolve, air pressure
(all optional)

Using Air to Move Water

Specific Concepts/Skills

1. Reinforce the concept that air and water cannot be in the same place at the same time.
2. Air may displace water, and water may displace air.

Note to Teacher

During this lesson you will demonstrate how to transfer air from an air-filled container to a water-filled container. Practice this before introducing the lesson.

Fill one cup with water by sinking it in the MPC. Lower the second cup, mouth first, into the MPC so that it remains filled with air. You are now going to transfer the air from the air-filled cup to the water-filled cup. Raise the water-filled cup up so that three fourths of it is out of water (see illustration). Move the air-filled cup under the raised water cup and tip the air cup slightly. Bubbles of air will leave the tipped cup and enter the water-filled cup. This air will force water out of the water cup. As you continue to tip the air-filled cup, it will fill with water. The bubbles leaving it will collect in the water cup. The use of colored water will make the investigation more obvious.

The Activity

1. Distribute the materials.
2. Review the previous lesson by first demonstrating what the youngsters did. Ask for an explanation of why the inside of the cup did not get wet when the cup was placed straight down and why it did not get wet when the cup was tipped.
3. Say: "We observed that as the cup was tipped, water entered the cup and air bubbles left. During this investigation you are going to try to capture the air bubbles that leave the cup."
4. Demonstrate what each team is to do. (For more advanced students you may pose the problem: How can you capture the air bubbles that escape from the tipped cup? Allow them to find their own solutions.)
5. Say: "Now, working together with your partner, try to repeat what you saw."
6. Allow the youngsters time to work out the necessary technique. You may have to help those with less coordination.
7. Review by asking:
   a. "What left the air-filled cup?" (Bubbles of air.)
   b. "Where did the air bubbles go?" (They entered the water and rose up into the water-filled cup.)
   c. "What happened when the air bubbles entered the water-filled cup?" (They pushed the water out of the cup.)
   d. "What do you think the bubbles were made of?" (Air, since the water cup is now filled with air. If some doubt this, place the cup mouth down into the water. It will not fill with water, showing it contains something—air.)
   e. "Can both air and water be in the same place in the cup at the same time?" (No.) "How do you know?" (The air forced the water out of the cup.) However, some youngsters might recall that fish get air from water and therefore air and water are in the same place. Point out that air is dissolved in

water. But, very little of the air in the cup can dissolve in the water and they therefore remain separate.

f. The following question may be raised by the youngsters but the answer, at this time, may prove to be too difficult to understand: ''When you raise the water-filled cup partially out of water, why doesn't the water fall out of the cup?'' (Air pressure keeps the water in.) Give the youngsters a chance to develop their own hypothesis. If collectively or individually they arrive at a satisfactory answer, fine; if not, they have an unanswered question and life is full of them. Lesson 1-10 involves the concept of air pressure and you might want to use it following this lesson.

Investigations at Home

Ask the youngsters to repeat the investigation at home using larger containers and with the assistance of someone at home. (Parents love to get involved in homework they understand and can enjoy. The youngsters will get a kick out of teaching science to Mom or Dad.)

Give each youngster a straw. Ask them to find at least two ways to replace the water in a cup with air using a straw. (Blow in an inverted cup filled with water. Drink the water and air replaces it.)

Evaluation

1. Were all youngsters able to manipulate the cup and transfer the air?
2. Do most youngsters realize that air is something and that air and water can't be in the same place at the same time?
3. Was the environment of the ''laboratory'' conducive to careful investigation and observation?
4. Are you and the custodian still speaking?

LEARNING SPECTRUM

1390 Westridge Drive
Portola Valley, CA. 94025
(415) 851-7871

THE NEW MICROSCOPE LABORATORY AND ACTIVITY BOOK

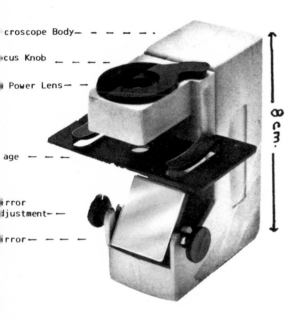

Microscope Body
Focus Knob
Power Lens
Stage
Mirror Adjustment
Mirror

The Science on a Shoestring (SOS) microscope has excellent optics, a sturdy body, is maintenance free, guaranteed, and costs only $5.00. It comes with a 30-power lens that is particularly easy for beginners to use. An optional 100-power lens is available that provides magnification for more advanced study. The SOS microscope may be used to observe regular slides and opaque objects. It can also serve as a mini micro-projector. Its small size makes it an ideal field instrument. The SOS self-guiding Microscope Activity Book provides an ideal way to introduce students to the microscopic world and to generate interest in science fair projects.

The SOS DeLuxe or Regular Laboratory Package is a convenient and inexpensive way to obtain SOS microscopes and all supportive materials.

DeLuxe Lab Quantity	Regular Lab Quantity	Description	Cost Each	Cost DeLuxe Lab	Cost Regular Lab
30	15	SOS Student Microscope with 30-power lens	$ 5.00	$150.00	$ 75.00
30	15	100-power Compound Lens	1.50	45.00	22.50
144	72	Microscope Slide	.10	14.00	7.20
30	15	Microscopy Depression Slide Plastic	.25	7.50	3.75
2 oz.	1 oz.	Cover Slips	4.00	8.00	4.00
30	15	Dropper	.25	7.50	3.75
30	30	Magnifying Lenses (5-power)	.30	9.00	9.00
15	5	SOS Microscope Activity Book	5.50	82.50	27.50
1	1	Microscope Repair Kit	1.25	1.25	1.25
2 vials	1 vial	Alum Crystals	1.25	2.50	1.25
50	50	Small Letter "e"	.50	.50	.50
2 bottles	1 bottle	Cell Stain	2.50	5.00	2.50
1	1	Teacher's Edition, Microscope and Activity Book	8.00	8.00	8.00
2	1	Brine Shrimp Mix	2.00	4.00	2.00
		TOTAL COST		$345.15	$167.95
		SOS TEACHER DISCOUNT PRICE		$285.00	$150.00

Free Bonus: Order a Deluxe Lab before December 1985 and receive a brine shrimp hatchery which includes an electric pump and everything necessary to grow brine shrimp, and a penlight that converts the SOS microscope to a microprojector.
- Order a Regular Lab and receive a free penlight.

MICROSCOPE SAMPLING KIT including 1 SOS microscope with 30-power and 100-power lenses, 5 slides and coverslips, and the SOS Microscope Activity Book: $15.00

PLEASE TURN OVER FOR ORDER FORM

LEARNING SPECTRUM

1390 Westridge Driv
Portola Valley, CA. 9402
(415) 851-78

1985 ORDER FORM FOR SCIENCE ON A SHOESTRING (SOS)
MICROSCOPES AND MICROSCOPE LABORATORIES

	Quantity	Total Amount

1. Deluxe SOS Microscope Laboratory @ $285.00
    (including bonus penlight and complete brine shrimp factory)

_____

2. Regular SOS Microscope Laboratory @ $150.00
    (including bonus penlight)

_____

3. SOS Microscope with 30-power lens @ $5.00 each
    10 for $45.00

_____

4. SOS 100-power lens @ $1.50 each

_____

5. Microscope Sampling Kit @ $15.00 each

_____

    Handling and Shipping: $6.00 per
    Laboratory . . . . _____

    Handling and Shipping: $.50 per
    Microscope to a maximum of $5.00 . . . . _____

    Handling and Shipping: $1.00 per
    Microscope Sampling Kit . . . . _____

    Sales Tax (California only) 6-1/2% . . . . _____

                              Total Amount . . . . _____

Ship to: Bill to:

_____   _____
School Name   Name

_____   _____
Street Address   Address

_____   _____
City, State, Zip Code   City, State, Zip Code

_____   _____
Attention   Purchase Order Number (Purchase
                                 orders accepted from educational
Mail to: Learning Spectrum   institutions only)
          1390 Westridge
          Portola Valley, CA 94025   Our Guarantee
Sorry, no purchase orders accepted   If you are not completely
for under $25.00.   satisfied, return the merchandise
Prices subject to change.   in its orginial container within 10
                                     days, and we will refund your money.

LEARNING SPECTRUM

1390 Westridge Drive
Portola Valley, CA. 94025
(415) 851-7871

T H E   N E W   M I C R O S C O P E   L A B O R A T O R Y   A N D   A C T I V I T Y   B O O K

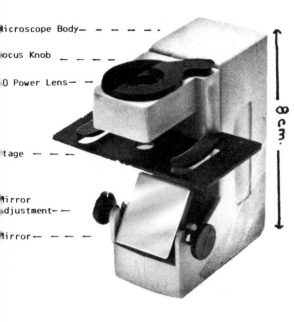

Microscope Body— — — — —
Focus Knob — — — — —
30 Power Lens— —
Stage — — —
Mirror Adjustment— —
Mirror— — — —

3.8 cm

The Science on a Shoestring (SOS) microscope has excellent optics, a sturdy body, is maintenance free, guaranteed, and costs only $5.00. It comes with a 30-power lens that is particularly easy for beginners to use. An optional 100-power lens is available that provides magnification for more advanced study. The SOS microscope may be used to observe regular slides and opaque objects. It can also serve as a mini micro-projector. Its small size makes it an ideal field instrument. The SOS self-guiding Microscope Activity Book provides an ideal way to introduce students to the microscopic world and to generate interest in science fair projects.

The SOS DeLuxe or Regular Laboratory Package is a convenient and inexpensive way to obtain SOS microscopes and all supportive materials.

DeLuxe Lab Quantity	Regular Lab Quantity	Description	Cost Each	Cost DeLuxe Lab	Cost Regular Lab
30	15	SOS Student Microscope with 30-power lens	$ 5.00	$150.00	$ 75.00
30	15	100-power Compound Lens	1.50	45.00	22.50
144	72	Microscope Slide	.10	14.00	7.20
30	15	Microscopy Depression Slide Plastic	.25	7.50	3.75
2 oz.	1 oz.	Cover Slips	4.00	8.00	4.00
30	15	Dropper	.25	7.50	3.75
30	30	Magnifying Lenses (5-power)	.30	9.00	9.00
15	5	SOS Microscope Activity Book	5.50	82.50	27.50
1	1	Microscope Repair Kit	1.25	1.25	1.25
2 vials	1 vial	Alum Crystals	1.25	2.50	1.25
50	50	Small Letter "e"	.50	.50	.50
2 bottles	1 bottle	Cell Stain	2.50	5.00	2.50
1	1	Teacher's Edition, Microscope and Activity Book	8.00	8.00	8.00
2	1	Brine Shrimp Mix	2.00	4.00	2.00
		TOTAL COST		$345.15	$167.95
		SOS TEACHER DISCOUNT PRICE		$285.00	$150.00

Free Bonus: Order a Deluxe Lab before December 1985 and receive a brine shrimp hatchery which includes an electric pump and everything necessary to grow brine shrimp, and a penlight that converts the SOS microscope to a microprojector.
- Order a Regular Lab and receive a free penlight.

MICROSCOPE SAMPLING KIT including 1 SOS microscope with 30-power and 100-power lenses, 5 slides and coverslips, and the SOS Microscope Activity Book: $15.00

PLEASE TURN OVER FOR ORDER FORM

LEARNING SPECTRUM

1390 Westridge Driv
Portola Valley, CA. 9402
(415) 851-787

1985 ORDER FORM FOR SCIENCE ON A SHOESTRING (SOS)
MICROSCOPES AND MICROSCOPE LABORATORIES

	Quantity	Total Amount
1. Deluxe SOS Microscope Laboratory @ $285.00 (including bonus penlight and complete brine shrimp factory)		
2. Regular SOS Microscope Laboratory @ $150.00 (including bonus penlight)		
3. SOS Microscope with 30-power lens @ $5.00 each 10 for $45.00		
4. SOS 100-power lens @ $1.50 each		
5. Microscope Sampling Kit @ $15.00 each		

Handling and Shipping: $6.00 per
Laboratory . . . . _____

Handling and Shipping: $.50 per
Microscope to a maximum of $5.00 . . . . _____

Handling and Shipping: $1.00 per
Microscope Sampling Kit . . . . _____

Sales Tax (California only) 6-1/2% . . . . _____

Total Amount . . . . . . _____

Ship to: Bill to:

_____   _____
School Name   Name

_____   _____
Street Address   Address

_____   _____
City, State, Zip Code   City, State, Zip Code

_____   _____
Attention   Purchase Order Number (Purchase
                                orders accepted from educational
Mail to: Learning Spectrum   institutions only)
          1390 Westridge
          Portola Valley, CA 94025   Our Guarantee
Sorry, no purchase orders accepted   If you are not completely
for under $25.00.   satisfied, return the merchandise
Prices subject to change.   in its orginial container within 10
                                      days, and we will refund your money.

MATERIALS
Per class
food coloring
Per team
1 MPC
sponge
Per student
1 clear plastic cup
1 pre-cut cork or Styrofoam
  hull (see illustration on
  next page)
2 pre-cut sails
2-3 pins

VOCABULARY
sail, air pollution, wind,
keel (optional)

Keeping a Sailboat Dry Under Water

Specific Concepts/Skills

1. Air is all around us. It is an important part of our environment.
2. Reinforce the concept that air and water cannot be in the same place at the same time.

Note to Teacher

This investigation is great fun. It also allows the youngsters to apply what they have learned by predicting if their submerged sailboats will get wet.

The Activity

1. Review previous lesson. Hold up a clear plastic cup and say: "We have learned that air is in the cup. Where did the air in the cup come from?" (It's in the room.)
2. Ask: "Can you see the air in the room?" (No.) "Can you feel the air?" (Have the youngsters move their hands back and forth quickly and they will feel a breeze made by moving air.)
3. Ask: "What do we call moving air?" (Wind, breeze, etc.)
4. Continue discussion of air in general with questions such as:

   a. "Can our air become dirty (polluted)?"
   b. "What causes air pollution?" (Car exhaust, burning leaves, industrial waste, etc.)
   c. "Can air move fast enough to lift a person off the ground?" (Yes.)
   d. "How high above the earth is air?" (Over 100 miles.)
   e. "What animals breathe air?" (Most.)

5. Say: "Clear your desks, and we will prepare for an investigation."
6. Pass out the materials listed and add food coloring to the water in the MPC. (You may want to have more mature youngsters cut out their own sails.)
7. Demonstrate how to make a sailboat and ask each youngster to build one (see illustration). Caution them not to get their sails wet.

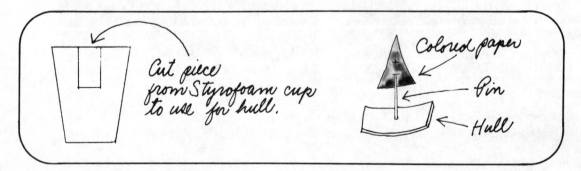

8. Ask: "If you place your sailboat on the water, do you think it will float?" Very carefully try it. "Would your boat and sail get wet if you were to put your cup over

it and then push your cup to the bottom of the MPC?'' (Allow the youngsters to give their hypothesis.) The hypotheses give an indication of the youngsters' understanding of the previous lessons.

9. Say: "Now carefully try it. Hold your cup down for a minute. Raise the cup slowly straight up and out of the water."
10. Ask: "Observe the sail. Is it wet or dry?" (Dry, unless the boat tipped over.) "Why is the sail dry?" (The air in the cup kept the water out of it.)
11. Say: "Try the investigation again, but this time look down through the cup. What do you see in the cup where the boat is?" (There is no color and therefore no water in the cup.)

Additional Investigations

12. Say: "Some boats tip over when the cup is raised. Try to build a sailboat that won't tip as easily and test it by blowing on the boat." (In one class a second grader added two pins to the bottom of the boat and found that it didn't tip. She "discovered" the keel.)

Ask: "Each time the cup is raised the boat tends to do something. Has anyone noticed what it does?" (It moves to one side of the cup.) "Can someone figure out why it does it?" (Allow the class first to make observations on their own, and then investigate until they form a hypothesis. Encourage them to test their hypothesis. Careful observation will reveal that water forms a convex bow in the sunken cup. The boat slides down the bow to the cup's edge.

Investigations at Home

1. Give them the boat and pins and have them repeat the investigation at home. Encourage them to teach someone at home what they have learned about air.
2. Have them locate as many things as they can that may pollute the air in their neighborhood.
3. Have them draw a picture of what might happen on the windiest day of all time.

Evaluation

1. Did two (or more) of the class predict the boat would remain dry?
2. Did the youngsters observe the boat's movement toward the cup or did you have to point it out?
3. Do the youngsters indicate an awareness of their environment when they are not studying science?
4. Do they still look forward to science investigations? Do you?

MATERIALS
Per team
1 MPC
1 clear plastic cup
1 piece of binder or
  construction paper (large
  enough to cover the mouth of
  cup)
1 sponge

VOCABULARY
volume (optional)

Using Air to Stop a Waterfall

Specific Concepts/Skills

1. Air has pressure
2. Air pressure can do work.
3. An observation should not be confused with a hypothesis.

Note to Teacher

This investigation is easy to perform and quite interesting, but a full understanding of why it works is difficult for young children to grasp. The investigation, therefore, is intended only as an introduction to the understanding of air pressure. Some teachers use it as a home investigation and stress observation rather than the concept of air pressure.

The Activity

1. Pass out the materials. Have the students fill their MPCs half full with water.
2. Say: "You are going to perform an investigation that is not difficult to do. Your job is to *observe* carefully what happens each time you do it and then try to figure out why it works."
3. Say: "Place the paper on top of the dry cup." (Demonstrate.) "What is in the cup?" (Air.) "Hold your hand on the paper and turn the cup over." (Demonstrate.) "Remove your hand and observe what happens." (The paper falls off.)
4. Allow the youngsters to describe their observations.
5. Say: "Now fill your cup about one half full of water. Put the paper on top. Hold the paper there and turn the cup over. Be sure the cup is above the MPC." (Demonstrate.) "Now carefully take your hand away from the paper."
6. Ask: "What do you observe?" (The paper remains on the cup and the water stays in the cup.)
7. Say: "Repeat what you did, but change the amount of water you use. Observe very carefully what happens. Repeat several times and observe what occurs."
8. Observations should include:

    If the paper is wrinkled, it won't stay on.
    The paper has to cover the mouth of the cup completely.
    Even if the cup is full, the paper stays on.
    The paper seems to be sucked in towards the cup (or is it being pushed in?). The observation should be that it is bowed inward. Why it is bowed in is a hypothesis. Take some time to explain the difference.
    Every time I turn the cup over, a little water leaks out of the cup. (*This observation is necessary to the understanding of why the paper stays on*. If no one observes it, you should hint at it.)

9. Ask: "Why do you think the paper stays on?" (Allow the youngsters to give their hypotheses. Encourage them to have their hypotheses agree with their observations.)

10. The answer: When the empty cup is turned over, the paper falls off because gravity pulls it down. However, when water is added to the cup and the cup is inverted, the paper bows inward and holds the water in the cup. What force could cause the paper to press against the cup? When the cup is first inverted, the air pressure outside the cup equals the air pressure within the cup, and the water in the cup starts to leave. This accounts for the few drops that *always* appear when the cup is first inverted. As the water leaves the cup, the level of water in the cup drops while the *volume*, or space, above the water increases. The amount of air above the water remains the same, but since the volume occupied by that air is greater, the air pressure decreases. So, a situation is created in which the outside air pressure is greater than the inside air pressure. As water continues to leave the cup, the *difference* in pressure increases until the outside air pressure slightly exceeds the total pressure of the air inside the cup, plus the water pressure, plus the weight of the paper. At this point the paper is pushed in and the escape of water is stopped.

For most younger children it would seem sufficient that they understand that air pressure *pushes* the paper in (rather than it being sucked in) and the paper holds the water in the cup.

Investigations at Home

Have the youngsters repeat the investigation at home using various-sized glasses and pieces of paper. Ask them to turn the cup in all directions to observe what happens. (The paper should hold the water in, since air pressure is the same in all directions.)

Evaluation

1. Were all children able to perform the investigation?
2. Are most youngsters improving their ability to observe and verbalize their observations?
3. Were the youngsters' hypotheses consistent with their observations?
4. The explanation is probably obscure to all youngsters. Were they able to accept that their answers were probably incomplete?
5. Was any youngster able to fully understand why the water stayed in the cup? If there is one, are you thinking about what you can do to help challenge her or his obvious talents?

MATERIALS
Per class
10 small containers (clear plastic cups or equivalent)
10 eye droppers
10 solutions (see sample result sheet, page 37)
Per team
1 MPC cover
masking tape
3 pieces red litmus paper
3 pieces blue litmus paper

VOCABULARY
acid, base, neutral, litmus paper, indicator

Investigating Acids and Bases Using Litmus Paper

Specific Concepts/Skills

1. Substances dissolved in water may form an acid, base, or neutral solution.
2. Litmus, an indicator, can be used to identify acids, bases, or neutral solutions.
3. Red litmus turns blue in a base. Blue litmus turns red in an acid. Red and blue litmus keep their color in neutral solutions.

Note to Teacher

This investigation introduces the youngster to some basic ways to determine whether a substance dissolved in water will form an acid, base, or neutral solution. Once the youngster understands the methods, he can apply them when describing the properties of a substance. The teacher should set up the 10 containers in advance. They should be numbered (using masking tape) for identification. The worksheet at right suggests what to use, but anything may be substituted that will provide samples of acids, bases, and neutral substances.

Lesson 1-12 has the youngsters repeat the test using red cabbage water instead of litmus paper. The same investigation result sheet may be used in 1-11 and 1-12.

Test for acid, base, or neutral with litmus paper.

The Activity

1. Say (modify as appropriate): "We can help identify the properties of a substance dissolved in water by testing the solution to learn if it is *acid, base,* or *neutral.* To test the substance we use an *indicator.* Today we are going to use the indicator called *litmus paper.* Later we will make our own indicator."
2. Review the meaning of *properties* as determined by the five senses. It is difficult to find out if a substance has the property of an acid, base, or is neutral by just using the five senses (and it may be dangerous). The indicator is a tool that will help us investigate those properties.
3. Distribute the MPC cover, two strips of masking tape about 20 centimeters (8 inches) long, and three or more pieces of red and blue litmus paper.
4. Demonstrate how to prepare the "testing board" (see illustration).
5. Proceed in one of the following ways, depending on the maturity of the youngsters.

INVESTIGATION RESULTS   Name _____

number	red litmus	blue litmus	cabbage juice*	substance	acid	base	neutral
1				lemon juice	X		
2				bleach-water		X	
3				water			X
4				aspirin-water	X		
5				milk			X
6				ammonia-water		X	
7				tea (concentrated)	X		
8				vinegar-water	X		
9				soap-water		X	
10				your saliva			

*Use this column if Lesson 1-12 is going to be used by the class.

Less mature: Demonstrate how to use the litmus paper. Place three clear glasses taped "acid," "base," and "neutral" before the class. Place the end of the red litmus paper in the acid solution (see sample result sheet for suggested acids and bases). Ask the class what change has occurred. (None.) Then dip the blue litmus. The class will observe that it turns red. Repeat using the base and neutral solutions. Class should conclude:

> Blue litmus turns red in an acid.
> Red litmus remains red in an acid.
> Blue litmus remains blue in a base.
> Red litmus turns blue in a base.
> Red and blue litmus keep their color in a neutral solution.

An easy way to remember this is: If the litmus turns *blue*, the liquid is a *base*. If the litmus turns red, the liquid is an acid.

More mature: Tell the youngsters solution 1 is acid, 2 is base, and 3 is neutral. Using that information, have *them* determine how the indicator may be used.

6. Place the ten numbered containers on a work area. Place an eye dropper in each. Have one member of each team place a drop or two of numbered solution next to the corresponding number on his testing board. Caution them not to mix up the droppers or the solutions will be contaminated.
7. Demonstrate how to test a drop. Tear off a small piece of litmus and dip it in the drop. Observe if there is a color change and note on the result sheet. Saliva may be tested by placing a small piece of litmus on the tongue.
8. Walk around and observe the youngsters at work, but don't give away answers. Youngsters enjoy testing the unknowns and should be allowed to develop their ability to test, observe, and record on their own.
9. Review by having youngsters discuss the results of their tests.

Investigations at Home

Give each youngster one piece of red and blue litmus. Ask them to test as many things as they can at home. Have them record the results on an investigation result sheet.

Evaluation

1. Did the team members cooperate?
2. After the initial confusion, did most youngsters settle down to work and accurately determine the unknowns?
3. Can all students satisfactorily perform the litmus and acid-base-neutral property test?

MATERIALS
Per team
pieces of red cabbage
2 clear plastic cups
1 eye dropper
10 toothpicks
1 MPC cover as prepared for
  1-11
Per class
very hot water

Making a Cabbage Water Acid - Base Indicator

Specific Concepts/Skills

1. The dye obtained from red cabbage may be used as an acid-base indicator.
2. In an acid cabbage water turns red. In a base it is blue.

Note to Teacher

Red cabbage water is an easily prepared acid-base indicator. If red cabbage is available this lesson should be used after Lesson 1-11.

The Activity

1. Cut the red cabbage into small pieces. Red cabbage is actually violet.
2. Have each team fill half a plastic cup with cabbage. Then cover the cabbage with very hot water.
3. The pigment in the cabbage will dissolve in the water, coloring it blue.
4. When the water is deep blue, have the youngsters transfer the water to a second plastic cup. (Cabbage may be cooked and eaten later if you like cabbage.)
5. The cabbage pigment is an indicator. In acid it turns red. In base it is blue. However, allow the youngsters to discover this on their own through investigation as in 1-11. Use the eye dropper to add the indicator to the solutions. A toothpick can be used to mix them. Place the results on the investigation result sheet.
6. You may wish to save the indicator for use in investigations requiring colored water.

Investigations at Home

1. The indicator may be taken home to test various common substances and the results reported back to class.
2. A volunteer may report on the uses of acids and bases.
3. Have a volunteer use the indicator to dye white cloth. How well does it work? Why isn't it used as a dye? (It changes color in acidic or basic solutions.)

Evaluation

1. Did the students exhibit increasing maturity when performing the test?
2. Did someone drink the indicator? It's not bad.
3. Are the youngsters looking forward to science time? And you?

Investigating Alka-Seltzer

Specific Concepts/Skills

The properties of a substance may be observed through careful investigation.

MATERIALS
Per team
1 Alka-Seltzer tablet
1 small test tube and test-tube holder
2 clear plastic cups
1 hand lens
match and/or candle
matches
litmus paper (optional)
candle (optional)

The Activity

1. Say: "Today we are going to investigate the properties of Alka-Seltzer."
2. Review the meaning of the term *properties*. List on the board all properties the class suggests, *e.g.,*

   a. color
   b. weight (heavy or light for its size; technically known as density)
   c. does or does not dissolve in water
   d. hard or soft
   e. smell
   f. feel
   g. taste (only if materials are known to be harmless)

3. Pass out the materials. Caution the class that they can have only one Alka-Seltzer tablet, and since they will be performing various tests, they should break it into *quarters* (review necessary math).
4. The following investigations are recommended. However, more mature youngsters may be encouraged to set up their own investigations.

Investigation I

a. Fill test tube one fourth full with water.
b. Drop in one fourth of an Alka-Seltzer tablet and observe.
c. When Alka-Seltzer is about half dissolved, cover the test tube opening tightly with your finger or palm. Wait a few moments. Describe what you feel. (Gas pressure.)
d. Hold the bottom of the test tube—describe what you feel. (It gets colder. The reaction of the Alka-Seltzer with water requires heat energy. The energy is obtained from the water, and consequently the water cools.)

Investigation II

a. Crush one quarter of the Alka-Seltzer using a test-tube holder. Observe the pieces with the hand lens. Are all the parts the same? Do you find any crystals? (The tablet is composed of a powder and crystals. The actual ingredients are listed on the wrap.)

Investigation III

a. Place one fourth of the Alka-Seltzer tablet in a clear plastic cup.
b. Add enough water to cover the tablet. Wait until the tablet is just about dissolved and then place a burning match or burning candle into the cup. Do not let the match touch the liquid or foam. Describe what occurs. Repeat again after 2 minutes. What happened? Why? (Carbon dioxide [CO_2] gas is produced when the tablet is in water. The escaping gas causes the effervescence. The CO_2 puts

out the match. After a short while the gas dissipates; the second match will burn.)

Investigation IV

Use the last one fourth of the Alka-Seltzer tablet to investigate any properties you wish. Be sure to write down what you are investigating and the results. (If this lesson follows 1-11 or 1-12, let us hope the youngsters request litmus or cabbage juice to find out if the Alka-Seltzer is acid base or neutral; it is acid.)

5. Have each team discuss the properties they discovered during Investigation IV.

Investigations at Home

1. Ask each youngster to select a substance, and make a list of its properties. In class they should read the list one item at a time to see if anyone can determine what their substance was. Later the substance should be shown to the class to compare the list of properties with the item.
2. Alka-Seltzer is taken to overcome several problems. Ask the youngsters to predict what each ingredient is supposed to do.

Evaluation

1. Did the youngsters discover properties you overlooked?
2. Did you need to take a couple of tablets when the lesson was over?
3. Are the youngsters improving their methods of investigation?

THEME 2

Change

Changes are occurring everywhere and at all times. However, unless they are sudden or irregular, they are often overlooked. The investigations grouped under this theme are designed to help the youngsters become aware of the changes occurring in themselves and their environment.

Some of the changes the youngsters will investigate are beyond their control. However, these changes can be carefully observed, measured, and data can be collected. By studying the data, youngsters may discover patterns that can make future change predictable. Investigations of weather, landscape, apparent solar movement, pollution, and growth fall into this category.

Most of the changes studied here, however, are initiated and to some extent controlled by the youngsters. This encourages the development of scientific skills and the concept of change, while recognizing that the average youngster needs to make progress in steps he can handle.

MATERIALS
Per class
(optional)
Polaroid camera and film

VOCABULARY
change, evidence, control

Observing Changes in the Environment

Specific Concepts/Skills

1. Our environment is continually changing.
2. We can effect some changes; others may be understood, but are beyond our control.

Note to Teacher

The lessons in Unit 2 are designed to focus the youngsters' attention upon the changes occurring within their environment. Students will investigate some changes that can be observed, measured, and perhaps explained but remain unalterable. Other investigations will emphasize the youngsters' role in initiating or modifying change. Lessons 1-1 and 1-1M should be presented as an introduction to change. The subsequent lessons may be used at any appropriate time and in any order.

The Activity

1. Initiate a discussion of change. Ask: "What changes have taken place or are taking place in this classroom?" (Give the youngsters plenty of time to think and find examples of change.) Some typical answers:

   The clock is moving.
   The sounds are changing.
   I'm getting hungry and I wasn't before.
   The desks are marked and scratched and once they weren't.
   The paint is dry but it was put on wet.
   The teacher's changed; she's gotten stricter.
   The floor polish is worn near the door.
   My pencil is shorter and chewed.

2.*Say: "Let's go outside and look for *evidence* of change. Each team is to find change and bring back to class evidence of that change. The evidence can be a drawing, the actual thing that is changing, a photograph (if Polaroid cameras are available), or a good written or oral description. Be sure to make note of where the change was observed." (For many classes the most effective way to conduct this investigation is to allow the students to work in small teams. The teams go outside and are given limits as to where they may walk, perhaps the entire school yard or even farther. The teacher is available to answer questions and supervise if needed. But it is essential that the youngsters look for and find evidence of change *on their own*. However, if the class is not mature enough to work independently, a group walk may be substituted.)

3. Have each team prepare a display of the changes (and evidence) they found. Some examples of change:

   a. Shadow moving (caused by sun's movement).

*This investigation is suggested in ESSENCE I (© American Geological Institute, 1970, 1971; published by Addison-Wesley in 1974.)

b. Temperature changing.

c. Ants moving things.

d. Leaf has fallen.

e. Leaf is changing color.

f. Yard was clean, now there is all kinds of junk.

g. Paint is peeling on the wall of the school.

h. Nails have rusted.

i. Clouds are moving.

j. Wall is cracked.

k. The soil has moved because of rain.

l. Air gets smoggy in the afternoon.

4. Some of the following questions may be used to extend the investigation:

a. What changes do you like?

b. What changes do you dislike?

c. What changes can you control, and what changes are beyond your control?

d. What changes would you like to make in your environment?

e. When trees are cut down, what changes may follow?

f. How has the invention of the automobile led to changes in our environment?

Investigations at Home

1. Say: "Look for evidence of change on your way home and in and around your house. Show your evidence and discuss the change with the class."

2. Ask the youngsters to bring in a photograph of themselves when they were younger. Mount the photographs (using corner mounts or masking tape rolled onto itself and placed on the back of the photo). Have the youngsters discuss what changes have occurred.

Evaluation

1. Did the teams find change? Were they able to bring to class evidence of the change?

2. Did the team members work maturely when sent outside to find change?

3. Did you allow the youngsters time to find change?

4. Do the youngsters show a greater awareness of change outside of science time?

> MATERIALS
> **Per class**
> Any box
> Data as suggested on
> next page

Building a Change Capsule

Note to Teacher

During this investigation, each youngster will build his own change capsule, and place a record of his characteristics within the capsule. Near the end of the school year, their capsule is opened and changes are measured and observed. The capsule may be made of any box. Decorations are optional.

What to Do

1. Discuss the purpose of a change capsule.
2. Discuss the kinds of data they could put into the capsule.
3. The data should be measurable so that comparison can be made near the end of the school year. For example:

   a. height and weight
   b. address
   c. length of foot
   e. fingerprint (use a stamping pad and white paper. Wipe off ink with a soapy paper towel.)
   f. written composition on food I like and don't like (the composition can be used to observe changes in handwriting and spelling as well as changing tastes).
   g. hair color (written, or better, a few strands cut from an identified location on the head).
   h. distance youngster can broad jump, number of pull ups, etc.
   i. seat in class.
   j. photo of themselves.
   k. whether you can rub stomach with one hand and tap head with the other (requires muscular coordination that improves with age).

4. Have each youngster build his own change capsule. Put the information within it, date it, and then seal it. Capsules may be stored where convenient and opened near the end of the year and comparisons made.

Growing Bird Seed

Specific Concepts/Skills

1. Different kinds of seeds grow into different kinds of plants.
2. Seeds are living things.
3. Young seed plants grow roots, stems, and leaves.

MATERIALS
Per team
1 MPC
1.5-cm (¼-in.) layer of cotton
 or 3 paper towels
masking tape
1 small envelope (for birdseed)
pinch of birdseed
Per student
1 hand lens

VOCABULARY
seed, root, stem, leaf

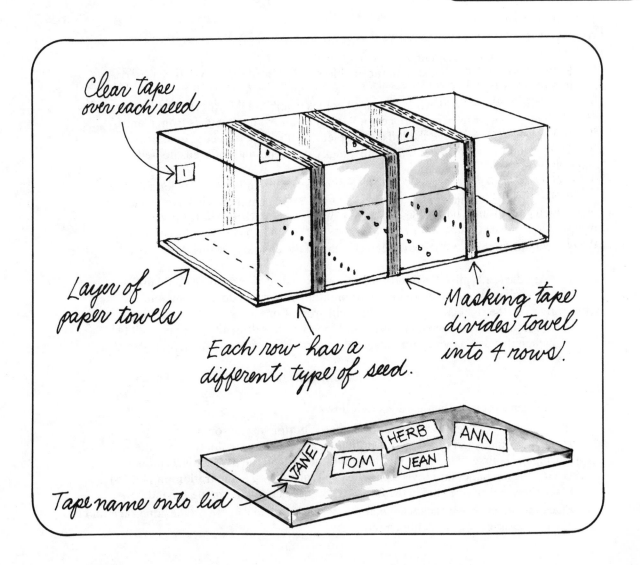

Clear tape over each seed

Layer of paper towels

Each row has a different type of seed.

Masking tape divides towel into 4 rows.

Tape name onto lid

VANE   TOM   HERB   JEAN   ANN

Note to Teacher

Birdseed, available in supermarkets, pet shops, and seed stores, is very cheap.
Although most birdseeds are untreated, some have been sterilized and will not grow.
This is usually noted on the commercial package or you can ask the sellers of seeds in
bulk. When in doubt, test a small amount. Growth usually begins in three-five days.
Most birdseed mixes consist of four-seven different kinds of seeds. Be sure the mix you
buy has at least three different kinds.

The Activity

1. Pass out one envelope containing a pinch of seeds to each team.
2. Ask: "Who knows what the stuff in the envelope is?" (Birdseed.) "What is it used
   for?" (Feed birds.)
3. Ask: "Does anyone know where birdseed comes from?" (The store.) "But, where
   does it come from before the store gets it?" (If they know that it comes from plants,
   fine; if not, help them to understand the problem, that is, where seeds come from.)
4. Ask: "Is your birdseed all the same? How many different kinds of seeds do you
   have?"
5. Tell the youngsters to separate the seeds into groups containing only one kind of
   seed. Each group should contain about 10 seeds. (Young children will have trouble
   sorting seeds. You may wish to have them plant the seeds without sorting them
   first.)
6. Tell the children to place the extra seeds back into the envelope. Collect the extra
   seeds.
7. Ask: "Are the seeds alive?" (Encourage the class to discuss how they could find out
   if the seeds are alive. They don't move, or breathe, or seem to do anything, but if
   they are given the right environment, they will grow. For young children the seed's
   growth is sufficient evidence that it is alive. However, it is difficult to define what life
   is, and when it ends. Delving into this problem can be rewarding with mature
   youngsters.
8. Pass out an MPC, some paper towels (or cotton), and three or four strips of masking
   tape to each team.
9. Have the teams prepare their gardens as follows:

   a. Place a layer of paper towels (or cotton) on the bottom of the container.
   b. Have them add water to the towels until they are wet, but not floating in water.
      During the course of the investigation continue to keep the paper towels moist.
   c. Divide the container into rows using the masking tape (see illustration on page 51).
   d. Place the seeds of one kind in the first row, of the second kind in the second row, etc.
   e. Place the lid on. Have each team member write his or her name on masking tape and
      place it on the cover for identification.

10. When planting is complete, have the teams place their container in a convenient location. Tell the class they are to *observe* any changes that occur. The kinds of observations and discussion you will encourage will depend upon the age and maturity of your students. Some suggestions for different maturity levels follow:

Young Children
After growth begins (3-5 days)

a. Which kind of seed began to grow first?
b. Are all seeds of that kind growing? (Usually not.)
c. Are all the different kinds of seeds growing? (Usually not.)
d. What is growing out of the seed? (Stem and roots.)
e. Is the seed getting bigger? (Yes, it absorbs water and swells.)
f. Were all the seeds alive?
g. Do different kinds of seeds grow into different kinds of plants? (Yes.)
h. Will the birdseed continue to grow without being placed in soil? (They will for several weeks, then, because of lack of minerals, they will die.)
i. Help the youngsters identify roots, stems, and leaves. Discuss the job of each. (Roots anchor plant and take in water. Stem supports leaves and transports water to leaves and food from leaves. Leaves make food for the plant.)

For More Mature Youngsters

j. How can you measure the growth rate of each kind of plant day by day? (A good time to introduce graphing; see illustration below and on next page for two methods. Have the class measure over a period of two weeks.)

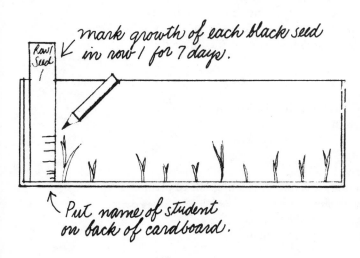

mark growth of each black seed in row 1 for 7 days.

Put name of student on back of cardboard.

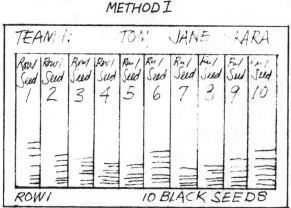

METHOD I

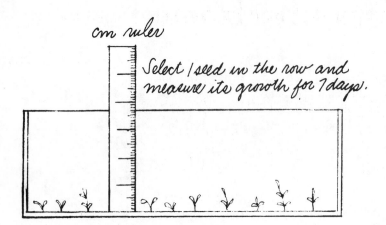

cm ruler

Select 1 seed in the row and measure its growth for 7 days.

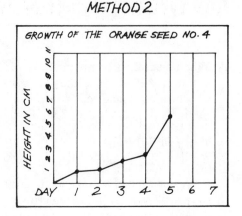

METHOD 2

GROWTH OF THE ORANGE SEED NO. 4

HEIGHT IN CM
1 2 3 4 5 6 7 8 9 10

DAY   1   2   3   4   5   6   7

k. Why do scientists often repeat their experiments many times before they reach a conclusion? In a discussion bring out that of the 10 seeds of any one kind they planted, one or two might not have grown. If by chance they selected only the seeds that wouldn't grow, they may have concluded that that kind of seed wouldn't grow. But, by using many seeds, they were able to come to a more accurate conclusion.)

l. What effect did room location have on seed growth? (Plants tend to grow toward the light; plant containers near heaters dry out faster [review evaporation].)

m. Sometimes mold may begin to grow in the container. This is great! The investigation is *not* ruined. Encourage the students to discuss what the mold is (a plant). Where did it come from? (Spores are in the air, and land everywhere, but usually the environment isn't right for growth. However, the environment in the container is good for some mold spores and they will grow along with the plants. A hand lens is very useful in seeing the parts of the mold.) A few students may want to investigate molds on their own and, of course, should be encouraged.

Investigations at Home

After several weeks of growth, the plants can be sent home for transplanting to soil. It is best to plant toweling and all so as not to disturb the roots.

Evaluation

1. Do all children know that different kinds of seeds grow into different kinds of plants?
2. Could they identify the roots, stems, and leaves of each plant?
3. Did their observations meet your expectations?
4. Did they want to try the investigation at home? Did you let them?

MATERIALS
Per student
fruit containing seeds

VOCABULARY
fruit

Growing Seeds Found at Home

What to Do

1. Ask: ''Where are seeds found?'' Guide student discussion. You may wish to cut an apple in half or open a grape to show seed locations.
2. Ask the children to bring in as many different kinds of seeds as they can. Ask them to bring at least two of each seed, one for show, and the other to plant and grow. When the seeds are brought to class, have the class try to identify what plant each seed comes from.

   Seeds usually available include tomato, cucumber, melons, apple, pear, orange, lemon, grapes, pumpkin, and pepper. Lima beans, pinto beans, corn kernels, nuts, grass seed, and many store-bought varieties of garden seed can be used.
3. Make a class seed chart. Have each child place one seed on the chart, and a drawing or magazine cutout of the parent plant.
4. Discuss with children that seeds found in different kinds of plants are different and will grow into plants like the parent plant.
5. Plant the seeds. Most seeds will germinate indoors and can be transplanted later into the ground. Pots can be empty milk cartons, egg boxes (one seed in each egg slot), the MPC, or plastic cups.
6. Review the problems in planting only one seed. Encourage discussion of how to improve changes of seed growth. (Proper environment, planting more than one seed, etc.)
7. Problems: Many seeds will take a long time to germinate and interest often fades after the planting is completed. One way to maintain interest, and keep the room jungle free is to have each youngster take his planting home. When growth begins, the plant can be returned to school and the spotlight of class attention turned to the ''farmer-scientist'' who demonstrates and discusses what has occurred.

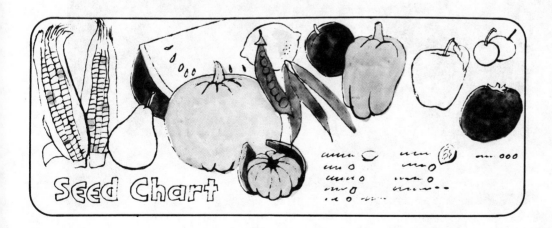

Seed Chart

MATERIALS

Per team

small pinches of alfalfa,
  mustard, and mung seeds
1 MPC
cotton to cover bottom of
  container (or paper towels)
3 strips masking tape
scissors

VOCABULARY

harvest, sow, germinate, health
food

Growing a Health Food Sandwich

Specific Concepts/Skills

1. Some seeds germinate quickly and the plants are good
   to eat.

Note to Teacher

This lesson is great fun and nutritious too. Youngsters will plant seed and within 7-10 days harvest and eat the results.

The Activity

1. Review by asking: "Are seeds alive?" (They may be.) "How can we tell if the seed is alive?" (Plant it.) "What kind of plant will grow from a radish seed? What kind of plant will grow from a grape seed? What environment do seeds need to *begin* to grow (germinate)?" (Moisture and a place for roots to grow.) "Why are seeds usually planted in soil?" (To obtain water and minerals from the soil.) "Will seeds grow on cotton?" (Yes, but later if they are to continue, they must be placed in soil.)

2. Say: "We have learned something about plants. Today we are going to plant three different kinds of seeds. Each seed will grow into a plant that we can eat and that is also good for us. After the plants grow, we will harvest and eat them."

3. Pass out the seeds. (You may wish to give a small pinch of each seed to each team, or package the seeds in small containers and distribute the containers.) The seeds selected are quite small. A small pinch should contain about 10-15 of each kind.

4. Pass out the MPCs and cotton (or paper towels).

5. To prepare the MPC for planting, have each team:

   a. Cover the bottom of MPC with cotton.
   b. Wet cotton.
   c. Use two strips of masking tape to divide the container.
   d. Write names of team members on a third piece of tape and tape to side of container for future identification.

6. Ask students to sow their seeds in rows, one row for each kind of seed. (Discuss how farmers sow seeds. Some youngsters may have helped their parents plant grass and should be encouraged to discuss what was done. Airplanes are used to sow seed over large areas and children may have seen pictures of this. For older children you might ask: "What does sowing your wild oats mean"? (Come to think of it, what *does* it mean?)

7. Finally, have the youngsters place the cover on the MPC loosely, allowing air to circulate through the container. (During this planting, we do not want to provide a good environment for mold. Low humidity in the container reduces chances of mold growth.)

8. Water as needed to keep cotton moist but not soggy.

9. Within 3-5 days, germination will occur. Discuss what the word *germination* describes. (The embryo plant beginning to grow.)

10. When plants are 10-15 centimeters (3-6 inches) high, they are ready for harvest. (Don't wait too long, as you want to maintain interest.)

The Harvest

Bring bread and mayonnaise (or some other spread) to class. Distribute scissors and explain how they can harvest their plants (cut at start of roots). Have youngsters harvest their crops and make health food sandwiches.

Investigations at Home

1. Children may wish to repeat this investigation at home and harvest the crops for their family. Let them. It's one lesson everyone can do right and thus earn the admiration and respect that we all need.
2. Ask youngsters to bring "health foods" to class and show and tell (and give out) the food they bring.

Evaluation

1. Were children able to establish a garden?
2. Was the harvest fun?
3. Do youngsters use the words *harvest, sow,* and *germinate* comfortably?
4. Did they want to repeat the investigation? (Don't do it—go on to the next lesson.)

Dissecting a Lima Bean Seed

Specific Concepts/Skills

1. A seed contains an embryo (baby plant).
2. A seed contains food to help the embryo grow.
3. When a seed germinates, the stem usually grows upward and the roots grow downward.

MATERIALS

Per team
1 clear plastic cup
cotton or blotter paper for
   inside of cup
lima beans (6 presoaked)
2 dry lima beans
masking tape
Per class (optional)
1 MPC
packaged soil

VOCABULARY
dissect, embryo, seed coat

Note to Teacher

The youngsters will study the lima bean seed. It is used because it may be dissected easily. Prepare for the lesson by presoaking six beans for each team one day in advance. It is a good idea to have a student do this for you.

The Activity

1. Ask: "What kind of seed is this?" (A lima bean.)
2. Say: "Today we are going to carefully *dissect* a seed." (Explain what *dissect* means—to cut apart in order to examine the parts.) "Why do you think we soaked the seed overnight?" (The seed takes in water and is easier to dissect.)
3. Say: "I would like one member of each team to come up and collect the materials needed for today's lesson. You will need (write on board if children can read) one cup, six soaked beans, two dry beans, and some cotton or paper towels (show approximate amount need). Please *quietly* come up for your materials."
4. Say: "Each 'scientist' is to dissect his own lima bean."
5. Directions and questions such as the following can help guide the lesson or you may wish to have them discover on their own.

   a. Carefully remove the seed coat.
   b. What do you think the seed needs the coat for?" (Protects the seed against injury and drying out.)
   c. Try to remove the seed coat from the dry seed. (It is difficult; that is why we soaked the seeds.) How else is the dry seed different from the wet? (It is smaller since the soaked seeds absorbed water.)
   d. Use your thumb nail to carefully split the seed. (The seed breaks into two even halves.)
   e. What does each half of the seed look like? (One half contains the embryo plant. Sometimes the seed will split and a part of the embryo will break off and stick to the other half as well.)
   f. Can you recognize the parts of the embryo plant? (Two leaves and another part, perhaps the stem or root.)
   g. What is most of the seed used for? (Most of the seed is food for the growing embryo.)

6. Say: "Let's further study the seed by planting it so we can observe its growth." Have the youngsters place the cotton (or blotter paper) around the inside of the cup. Then place the soaked seeds between the cotton and the inside of the cup. The seeds should be placed in different directions so that as the embryos grow the students can observe how the roots, stem, and leaves emerge and in what direction they grow. Add water to the bottom of the cup and allow the cotton to soak up the water.

7. Have each team print their names on masking tape to identify their container.
8. Examine the cotton each day to be sure it remains moist. Within a few days, growth will begin.
9. Have one team plant its lima beans in soil. Allow them to work on their own, using an MPC and packaged soil. Later they can "show and tell" to the class.
10. After growth begins (3-5 days), ask: "Which way do the roots grow? Which way do the stem and leaves grow? Does the position of the seed make a difference?" (No.)
11. For more mature youngsters, you may want to ask:

    "Why do the roots grow down and the stem up?" (The usual answer is: roots grow down to get water; the stem grows up so leaves will get light.) However, a perceptive student will realize the wet cotton surrounds the seed and, therefore, the roots don't have to grow down to get water. Another question might be, "Sure the leaves need light, but how does the plant know that? Can it think? (Plants have inborn responses triggered by their environment. They respond not only to light and water but to gravity and other stimuli as well. A more thorough discussion is available in many junior or senior high school biology books.)

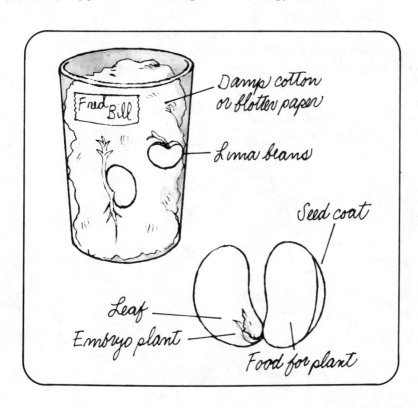

Investigations at Home

1. Give leftover beans to youngsters that want to plant them at home. Have them "show and tell" when they return the growing plant to school. (Differences will occur because of the various environments at home.)
2. Various experiments that compare one growth environment with another may be done by volunteers. For example, using milk cartons or any other container, plant several seeds in two identical containers filled with soil or cotton. Water both the same amount, but keep one container in the dark and the other in the light. Observe similarities and differences in growth.
3. Carefully dissect away a portion of the seed's food. Plant and later compare growth with undissected seed.
4. Encourage youngsters to develop their own investigations and supply them with needed materials for home or class.

Evaluation

1. Was the student attitude serious and thoughtful?
2. Were the youngsters able to place the seeds in varying positions against the cup?
3. Was their interest sustained between planting and first growth?
4. If the youngsters said they had planted lima beans 10 times before and knew all about them, did you:

   a. push on with the lesson?
   b. modify the lesson so as to add to what they knew?
   c. tell them to do it again—it's good for you?
   d. panic and spill the beans?

MATERIALS
Per team
1 small test tube
2 clothespins
1 votive or 10 cm (4'') candle
1 piece aluminum foil, about 10
  cm × 10 cm (4'' × 4'')
matches
1 eyedropper

Causing Water to Change its State

Specific Concepts/Skills

1. A liquid may be changed to a gas by adding heat.
2. A gas may change back to a liquid when the heat is removed.
3. Some experiments take time, patience, and cooperation.

Note to Teacher

In this lesson, youngsters heat water in a test tube. They observe the water changing from a liquid to a gas and perhaps (at the top of the tube) to a liquid again.

Carbon will collect on the outside of the test tube if it is placed in the candle flame. This will obscure the view of the boiling water and will lead to many interesting hypotheses. Frequently youngsters are convinced the test tube is burning. You may wish to encourage the class to discover what is really happening and why. (As the candle burns, some carbon in the wax does not combine with oxygen and is released into the air. When the test tube is placed in the flame, the carbon collects on it.)

A votive candle is less apt to tip over and therefore more suitable for young or clumsy students. The 10-centimeter candle is preferable for others.

The Activity

1. Review through discussion the nature of solids, liquids, and gases.
2. Say: "Today we are going to heat water in a test tube and observe what happens."
3. Review what the word *observe* means. Point out the difference between just looking at something and carefully observing it. Explain that, as scientists, they must very carefully observe what happens to the water when it is heated.
4. Demonstrate how they should use the candle. Place your votive candle on the aluminum foil and light it. (Note: If a taller candle is used, drip some wax in the center of the foil and place the candle on the liquid wax. When it solidifies, the candle will be held in place. Or place the candle on the foil and pinch the foil around the candle, as in the illustration.)
5. Pass out the student materials.
6. Have each team attach the clothespin to the test tube. A slight rotation of the test tube will help it slip into place.
7. Place a few drops of water in each test tube (or have them do it themselves).
8. Light each candle (or have them do it themselves). One method of having children light their own candle without using matches is for the teacher to light one child's candle. The person sitting next to that child places his or her candle wick into the flame. Each child in succession does the same until all candles are lighted.
9. Demonstrate how to heat the water in the test tube. *The test tube should not be pointed at anyone.* It should be moved to and fro to heat the water evenly. If held too close to the flame, the test tube will become coated with carbon. If this occurs, encourage the youngsters to *observe* this, along with what is happening to the water.
10. Say: "Carefully heat your test tubes and observe what happens." (If possible, students should write down their observations.)
11. Students should continue the investigation until no water is left.

12. Ask the following questions to encourage discussion during or after the experiment:
    a. What is happening to the water in the test tube? (It's boiling.)
    b. Do you notice anything leaving the test tube? (Steam.)
    c. What do you think it is? (Water.)
    d. Where is the water going? (Into the air.)
    e. Has the liquid water changed into a gas? (Yes.)
    f. How do you know? (The amount of water is decreasing.)
    g. Has anything happened to the test tube? (Bottom may have turned black.) Some youngsters may think the test tube burned. Have them touch the cool test tube with paper. The carbon will come off, and the clear glass will be exposed. Or place the warm test tube in a glass with cold water. The carbon will come off the test tube and float in the water.
    h. Where could the black material have come from? (The burning candle.)
13. Repeat the investigation. This time have one team member hold a small piece of aluminum foil over the boiling water, using a clothespin. (The foil may be torn from the candle holder or a new piece distributed.)
14. Have the youngsters lift the foil off the test tube.
15. Ask: "What is on the foil?" (Beads of water.) Where did it come from? (The hot water changed to steam and then back to water.)
16. Ask: "What caused the liquid to change to a gas?" (Heat.) "Why did the gas change back to a liquid?" (It cooled.)
17. Review the lesson and the concepts learned. That is, a liquid may be changed into a gas by adding heat. When the vapor cools, it may change back to a liquid.

Evaluation

1. Did the youngsters cooperate with each other?
2. Did they successfully complete the investigation?
3. Were their written or oral observations pertinent and accurate?
4. Did they understand why the liquid changed to a gas and back again?
5. Do you feel better about allowing them to use candles for a heat source?

Removing Water from Air

Specific Concepts/Skills

1. A gas may change to liquid when cooled.
2. Scientists make careful observations and hypotheses to explain their observations.
3. (optional) When water vapor cools, it may condense as water droplets.

MATERIALS

Per team
1 glass
3 or more ice cubes
1 container of liquid (milk, water, soda, etc.)
1 paper towel (to carry the ice in)

VOCABULARY

hypothesis, condense (optional), water vapor

Note to Teacher

During this lesson, youngsters will put ice in a glass and observe what happens. They will discuss their observations and hypotheses about the causes. Many youngsters may not fully understand why water collects outside the glass. Even so, they will profit from learning to *really see* what they have seen many times before.

The best way to carry out this investigation is to have each team perform it at its table or desk. However, if ice is difficult to obtain, the investigation may be modified and performed as a teacher demonstration or as a home investigation.

The Activity

1. Review the previous investigation. Have the youngsters discuss how heat was used to change a liquid (water) into a gas (water vapor).
2. Review why the water vapor changed back into a liquid. (The water vapor cooled near the top of the test tube.)
3. Say: "During this investigation you will do some things you have done before, and see some things you have seen before, but this time I want you to observe carefully what happens, and then give a hypothesis as to why it happened." Remind the class that a hypothesis is an intelligent guess.
4. Pass out the student materials.
5. Ask the youngsters to place the ice into the glass and then carefully add the liquid until the glass is about three fourths full.
6. Ask: "How many have done something like this before?"
7. Tell the class to observe carefully what they see happening. If the youngsters are capable, you may want them to write down their observations. Otherwise, they can draw what occurs and make mental notes. The students will observe many things that we are not directly concerned with. Don't discourage this. Rather, it is important that they be encouraged to make observations. However, the thing we are most concerned with here is the formation of water droplets on the *outside* of the container. When in the course of discussion this is mentioned, ask:
8. "Where did the liquid on the outside of the container come from?" Answers may include:

   It jumped through the glass.
   There are holes in the glass.
   I saw that before; it always happens.
   I spilled some on the outside.
   A good fairy puts it there.

10. As skillfully as you can, help the class to distinguish between their observations (water formed on the outside of the glass) and their hypothesis (why it formed).

11. Have the class discuss each hypothesis as it is mentioned and encourage testing of the hypothesis.
12. Most younger children (first through fourth grade) will discard a hypothesis that doesn't make sense, but will have trouble coming up with a workable hypothesis to explain the appearance of the water drops. Review the previous lesson and help the youngsters form a workable hypothesis. (The ice in the glass cooled the glass. The air that came in contact with the glass was cooled and the water vapor in the air changed into water, collecting as drops on the outside of the cup.)
13. Have the youngster taste the drops. They are pure water removed from the air.
14. At this point you may wish to introduce the words *condensation* and *condense*. When water vapor changes into water, it is said to condense. The water vapor in the air condensed on the outside of the glass.

Investigations at Home

1. Have the youngsters repeat the investigation at home. If possible, they may act as teacher and a parent, brother, or sister becomes the student. (Children get a kick out of knowing something the family doesn't and, as you know, one learns by teaching). In class they may discuss their home investigation.
2. Ask those who take a hot shower to observe the mirror in the bathroom. What happens to it and why? (Moisture in the air collects [condenses] on the mirror.)

Evaluation

1. Can the children distinguish between an observation and a hypothesis?
2. Are their observations accurate?
3. Do they accept what you say blindly or do they question?
4. Does more than half the class understand why the water drops collect on the outside of the glass?
5. Do you feel comfortable saying "I don't know?"

MATERIALS

Per class
strike-anywhere matches
safety matches
a glass
a piece of sandpaper
hand lens (optional)
cigarette lighter (optional)
flint and steel striker (optional)
solar heater (optional)

VOCABULARY
fuel, friction, ignite, safety
match, nitrogen, chain reaction

Investigating the Requirements for Fire: How Does a Match Work?

Specific Concepts/Skills

1. In order to burn a fire needs fuel, oxygen, and heat.
2. When two things rub against each other friction occurs and heat is released.
3. Fire can be dangerous and caution is needed when using it.
4. Some matches contain all the materials necessary to start a fire; others lack one ingredient.

Note to Teacher

Children who understand the causes and effects of fire are more likely to respect the potential benefits and hazards involved in using it. Should an emergency arise the informed youngster will be better prepared to respond intelligently.

This lesson is written mainly as a demonstration-discussion. However, it may easily be modified to include student investigation if you feel your class is capable.

The Activity

1. Ask: "What do you think is needed to make a fire?" (Write whatever they mention on the board, placing each in one of three categories of necessary items: fuel, oxygen, and heat. For example:

wood	air	matches
paper	oxygen	
gasoline		
gas		
stove gas		
hair		

2. Say: "All the things you mentioned in this column are known as *fuels*." (Write it above the column.) "They are things that will burn." Suggest additional items and ask if they belong in the fuel column (brick, iron, wool, paint, etc.).

3. Say: "The things in the second column are gases. Air is made of mainly two gases, oxygen and nitrogen. The nitrogen doesn't help things to burn. The oxygen does. Therefore only oxygen should appear in this column." (That is not strictly correct as a few other gases will support combustion such as chlorine, fluorine, etc.)

4. Say: "The match can supply the heat necessary to start the fire. Do you know of any other things that can supply heat to start a fire?" (Flint and steel can be demonstrated using a cigarette lighter or striker. The sun's energy can be used by focusing sunlight onto paper using a hand lens. The recommended student hand lens will work but *very* slowly in winter months. It would be better to use a lens of larger diameter. Small solar heaters costing about a dollar are available. They reflect light to a point where a cigarette is usually placed. Paper can be substituted and ignited.)

5. Distribute the sandpaper (optional)

How does a match work?

6. Ask: "We know a burning match can start a fire, but how does the match get started?" (It needs heat, fuel, and oxygen, too). "What do you think it uses for fuel?" (Hold up a nonsafety or strike-anywhere match or use an enlarged drawing.)

"It uses the cardboard or wood and its 'head'. Where does it get its oxygen from?" (The air.) "Where does it get the heat to get started?" (Allow discussion.)

7. Say: "Rub your fingers quickly against the palm of your hand. What do you feel?" (Heat.) "Now rub your fingers lightly against the piece of sandpaper." "Which made your finger warmer?" (Sandpaper.) "Why?" (It's rougher.)

8. Say: "When two things are rubbed together, *friction* occurs and heat is formed. The rougher the things are, the more friction and heat will form. When the match head is rubbed against something, friction occurs and the match head gets warm."

9. Rub the nonsafety match against a glass (or window pane). Ask: "Why doesn't it *ignite*?" (Not enough friction because the glass is smooth.) Rub it against the sandpaper. (It ignites.) Ask why.

10. Explain (modify the explanation for maturity level of class or individuals). A match contains the things necessary for a *chain reaction*. To start the reaction, the head is rubbed against something rough. Friction occurs and heat is released. The small amount of heat causes a chemical (a phosphorus compound) in the head to ignite and a small flame develops. This flame is hot enough to cause a second chemical (potassium chlorate) to release oxygen. The oxygen makes the flame hotter and causes a third chemical (sulfur) to begin burning. The burning sulfur gives the match its odor. Once the head is burning, the heat causes a wax coating to melt and it starts to burn. By then the fire is hot enough for wood or cardboard to burn on its own. The chemicals that helped the match get started are quickly consumed.

Strike match - Friction
↓
Heat
↓
Phosphorus ignites
↓
Potassium chlorate releases oxygen
↓
Flame gets hotter
↓
Sulfur ignites
↓
Wax melts and burns
↓
Flame gets hotter
↓
Match stick catches

11. Take out a safety match and strike it against glass. Then try it against sandpaper. Repeat. Ask: "Why won't this match ignite?" (Allow discussion.)
12. Explain: This is a safety match. It contains all the chemicals the strike-anywhere match has except one (the phosphorus compound). That chemical is the one that starts the chain reaction. That missing chemical is located on the match box or folder. (Demonstrate.) When the safety match head is rubbed against the chemical (mixed with a grit to increase friction), the chain reaction begins.
13. Review by asking:

    a. What do all fires need in order to burn? (Heat, fuel, oxygen.)
    b. How can you cause friction? (Rub two things together.)
    c. How can I find out if this is a safety match? (Rub it against a rough surface.)
    d. Why can it be dangerous to play with matches?
    e. How can you start a fire without matches? (Flint or steel spark ignites fuel; magnifying lens and sunlight, etc.)
    f. Broken soda bottles have been known to start forest fires. How? (The glass, particularly the bottom, can act as a magnifying lens and cause dry grass to ignite.)
    g. Can something catch fire by just being near a fire? (Yes. If the fire is hot enough, the heat may ignite a nearby substance.)
    h. Why are safety matches called safety matches? (They are hard to ignite accidentally.)
    i. Can a safety match be dangerous? (Yes.)

14. Conclude with a discussion of fire and the need for caution when using it.

Evaluation

1. Can all youngsters name the three things necessary for fire?
2. Do most have a feel for what a chain reaction is?
3. Can anyone describe the match chain reaction?
4. Have you encouraged youngsters to discuss friction in other contexts?

Investigating Ways to Extinguish Fire

Specific Concepts/Skills

A fire may be put out by removing its fuel, heat or oxygen.

MATERIALS

Per class
1 large teacher candle — for demonstration and relighting students candles
matches

Per team
1 small test tube
1 small candle
1 7-cm × 7-cm (3″ × 3″) square of aluminum foil
matches (optional)
eye dropper

VOCABULARY
extinguish (optional)

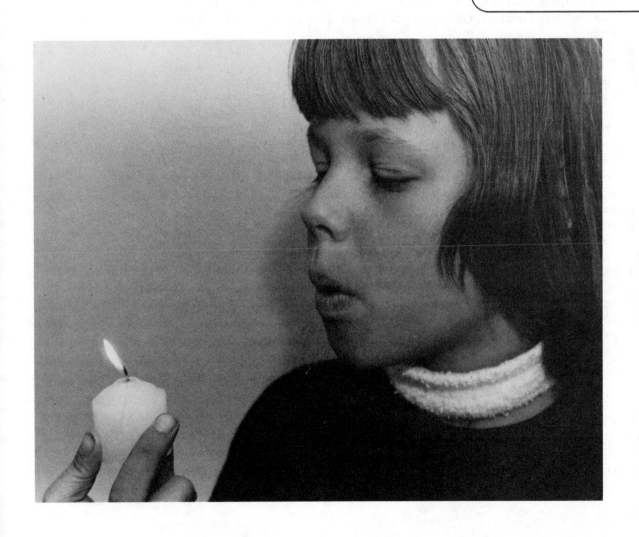

Note to Teacher

Much of the material used in this lesson is also used in Lesson 2-9. If you and your class are up to it you may want to combine the lessons

The Activity

1. Review the things necessary for fire. (Fuel, heat, and oxygen.)
2. Ask: "How do you think a fire can be put out?" (During the discussion bring out that each method involves *removing* one or more of the necessary ingredients for fire.
3. Say: "Clear your laboratory tables (desks) for investigation."
4. Pass out the student materials. Demonstrate the correct way to set up the candle (see illustration on page 63).
5. Light the student candles or have them do it.
6. Say: "Observe the candle flame. Describe what you see." (More advanced youngsters can draw or write down their observations. There are *many* things to be seen. Allow plenty of time for observation. Turning off room lights can help.)
7. Tell the children to lower their test tube mouth-side down over the flame.
8. Ask: "What happens to the flame?" (Goes out.) "Why did it go out?" (Bring out in discussion that the flame used up the oxygen in the test tube and because it needs oxygen to continue burning, it went out.)
9. Ask: "Why do you think a candle goes out when you blow on it?" (Allow discussion but direct it toward removal of heat, fuel, or oxygen.)
10. Have youngsters relight their candle (or you do it).
11. Say: "Carefully blow on your candle flame, but *don't blow it out*. Observe what happens to the flame."
12. Say: "Now blow out the flame, and try to explain why it went out." (Most should come up with good hypotheses related to the things necessary for fire. For example:

    a. You blow the fuel away. (True; the wax vapor is blown away.)
    b. You cool the flame. (True; you blow in fresh room-temperature air, which moves the hotter air away.)
    c. You blow away the oxygen. (No, you actually blow in fresher air and therefore add to the oxygen supply.)

13. A teacher demonstration or student investigation may be used here to explore what happens when the flame is blown out. Light a candle and then slowly blow it out. Quickly place a burning match near the wick. The flame will seem to jump from the

match to the wick. Why? When the flame is blown out, the fuel (wax in vapor state) is blown away from the wick and heat. The match ignites the wax vapor and the flame quickly moves back to the wick and a supply of new wax vapor. Allow the youngsters to discuss the demonstration or investigation.

14. Say: "Fill your test tube with water." (Use the sink or dip into a water-filled container.)
15. Relight the candles.
16. Say: "Carefully add water a drop at a time (using the eye dropper) to the candle flame. Observe what happens."
17. Ask: "Why do you think the water put out the fire?" (Water cools the fire and separates the fuel from the oxygen.)

Review

18. Review by asking questions such as:
    a. What do firemen usually use to put out fires? (Water.)
    b. Where do they get the water? (Carry it in their truck or fire hydrant.)
    c. How does the water put out the fire? (Cools fire, cuts off oxygen supply.)
    d. Sometimes a blanket of sand or salt is used to put out a small fire. Why do they work? (Cut off oxygen supply.)
    e. When you blow on a match, the flame goes out, but when you blow into a campfire, the flame gets bigger and hotter. Why? (Unlike the small match fire, you cannot blow away the fuel or the heat. You can only add a fresh supply of air [oxygen], which helps the fire burn more intensely.)
    f. Why isn't pure oxygen used in space capsules? (Among other reasons, fear that if a small electrical fire started it could quickly get out of control. This did happen, killing three astronauts.)
    g. What can you do to prevent an uncontrolled fire in your house? (Don't play with matches. Keep things you don't intend to burn away from open fire, etc.)
    h. If you put a burning match into boiling water, would it go out? Why?
    i. If you need the fire department how could you contact them?
    j. Why is a false alarm harmful to everyone?

Investigations at Home

Teacher "homework:" Try to arrange for the class to visit a firehouse or to have a firefighter visit the class.

Evaluation

1. Were the youngsters physically able to carry out the investigations and cooperate with their partners?
2. Were they able to understand that fire needs oxygen, heat, and fuel and that by depriving it of one or more of these components the fire will go out?
3. Did the youngsters exhibit increasing maturity in working with fire?
4. Could they relate what they learned through investigation to their home environment?

MATERIALS

Per class
food coloring
matches or candle to light
birthday candles
Per team
1 2-cm (¾'') candle
1 test tube to cover candle
1 wide-mouth container (or
  substitute)
matches

Investigating the Effect of Burning on Air Pressure

Specific Concepts/Skills

1. Oxygen is used during the burning process.
2. Air pressure can do work and overcome gravity.

Note to Teacher

During this investigation, the student will place a test tube over a burning candle resting in a cup with water. The candle will go out and the water will rise about one fifth of the way up the test tube. An interesting investigation, but the significance may be overlooked. The teacher can help more mature youngsters seek to understand the rather complicated chain of events that led to the water's overcoming of gravity.

Advanced Preparation

Cut 10 cm (4-inch) candles into approximately 2 cm (¾-inch) lengths. Flatten one end so it can stand, and free the wick from the other end.

The Activity

1. Review the previous lesson.
2. Pass out the materials. Have the youngsters place the candle in the center of their container. Pour the water into the container around the candle (see illustration at right).
3. Ask: "What is inside your test tube?" (Air.)
4. Direct the students to lower the inverted test tube over the unlit candle until the lip of the tube touches the water. Ask: "What did you observe?" Have them repeat, moving the test tube down faster. (Youngsters will observe various things and they should be discussed and looked for by other teams.)
5. Have the students light their candles (or you light them).
6. Say: "Lower your test tube quickly until it touches the water. Observe what happens."
7. Ask: "What happened this time?" (Flame went out; water rose up the test tube, carrying the candle with it.)
8. Ask: "Who would like to give a hypothesis to explain what happened?" (There was air in the test tube. Part of the air is oxygen [about 20 percent, or one fifth]. When the test tube was lowered into the water, burning candle quickly used up the oxygen in the test tube. Therefore, there was less gas and correspondingly less pressure inside the test tube. The air pressure outside remained the same and pushed the water up into the test tube. The water rose about one fifth of the way up the tube because about one fifth of the gas in the test tube was oxygen, and it was used in the fire.) The explanation may be too difficult for the majority of youngsters to grasp. But this surprising investigation and what they can understand of it will move them along towards internalizing the concept of air pressure and the chemistry of burning. So, don't be upset if many students can't really understand why it happened. They will remember *what* happened, be puzzled by it, and hopefully worry about it until they are ready to understand.

9. For more advanced students only: Have them repeat the investigation, lowering the test tube onto the burning candle at various speeds. Say: ''When the candle burns, it combines with oxygen to form carbon dioxide gas. This gas is then in the test tube. Why doesn't it increase the pressure in the test tube and prevent the water from rising? What effect does the volume of the candle have on the water level? What is the significance of the slight bubbling at the lip of the test tube?''

Investigations at Home

Investigation may be repeated at home *under parent supervision* using a saucer and glass and candle.

Evaluation

1. Were most youngsters curious as to why the water rose and the candle floated?
2. Did the team members cooperate during the investigation?
3. Did they listen to other youngsters' observations and hypotheses with interest and critique them?
4. Did they want to repeat the investigation to observe and ''find'' answers? Did you let them?
5. Do you feel more comfortable allowing the youngsters to use burning candles and investigate on their own?

MATERIALS

Per team

2 multipurpose wax cups

1 tablespoon baking soda

3 tablespoons vinegar

1 regular candle

1 7-cm × 7-cm (3″ × 3″)
 piece of aluminum foil

VOCABULARY

physical change, chemical
change

Making and Using Carbon Dioxide

Specific Concepts/Skills

1. Changes may be classified as physical or chemical. Physical changes involve changes in size and shape only. Chemical changes involve changing a substance into another kind of substance.
2. When vinegar and baking soda are mixed, a chemical change occurs and carbon dioxide (CO_2) gas is formed.
3. Carbon dioxide is colorless, odorless, will not burn, and is heavier than air.
4. Carbon dioxide is used to put out certain kinds of fires and as an effervescent in sodas.

Note to Teacher

This lesson is surprising and delightful. However, the concepts of physical and chemical change may be difficult for children (and teachers) to grasp. You may wish to go through the activity anyway, de-emphasizing these technical terms and concentrating instead on the concept of change and the process of observation.

The Activity

1. Begin by discussing change. You might tear some paper to show change in shape. Point out that even though the paper is smaller, it is still paper. The kind of change in which only shape or size is changed is known as a *physical change*.
2. Ask the children to mention other types of physical changes. (Making chocolate milk, sewing a sweater, boiling water, etc.)
3. Light a match. Ask: "Is the wooden stick changing?" (Yes, it's burning.) "After the wood burns, is it still wood?" (No.)
4. Say: "When a substance changes and is no longer the same substance, a *chemical change* has occurred. The wood has been changed to this black substance (carbon) and is no longer wood. Burning the wood caused a chemical change.
5. Ask the youngsters for other examples of chemical change (baking a cake, a plant growing, making food, etc.).
6. Say: "Today we are going to perform an investigation involving change. I would like you to decide if the change is physical or chemical."
7. Pass out two multipurpose wax cups and one candle to each team. Have a monitor pour about 3 tablespoons of vinegar in one cup, and one tablespoon of baking soda in the other.
8. Light the candles.
9. Ask the children to tip their candle into the cup containing baking soda and to observe any changes. They may observe that the flame points up, wax drips, etc. Have them repeat with the vinegar container. Advise not to touch the flame to the vinegar (or flame will go out).
10. Ask them to pour the vinegar into the cup with baking soda. Advise them not to lift the baking soda container off the table. (The CO_2 gas may escape.)
11. Ask: "What changes are occurring?" (Bubbling, smell may change.)
12. Say: "Now carefully tip your candle into the bubbling cup and observe what happens." (Flame goes out at once.)
13. Relight candles and allow them to try again.
14. Ask: "What could have caused the flame to go out?" (The vinegar and baking soda bubbled; the bubbling was caused when a gas formed; the gas could have put the flame out.)

One obvious change that occurs when vinegar is poured into the cup with baking soda is that both children are delighted with the result. A more subdued but equally gratifying reaction occurs when the children observe the effect of CO_2 on the candle, as at right.

15. Ask: "When the vinegar and baking soda were mixed, did a physical or chemical change occur?" (Vinegar is changed—the smell is gone; baking powder is also changed.)
16. After some discussion, tell them that the two chemicals (vinegar and baking soda) combined, making a chemical change. They formed the gas carbon dioxide. Scientists have a short way of writing carbon dioxide: CO_2. The C stands for carbon and the O_2 for oxygen.
17 Ask: 'Who can tell me something about CO_2?"

   Children may answer: puts out fire
   heavy (it stays in open cup)
   has no color
   no smell
18. Ask: "Does anyone know how this gas is used?" (Fire extinguishers, soda, etc.)
19. If you did Lesson 1-13, Investigating an Alka-Seltzer, you might discuss the similarities. The gas formed by Alka-Seltzer was also CO_2.

Investigations at Home

1. Encourage the youngsters to repeat the investigation at home under parent supervision. Ask them to explain in class what they did at home.
2. A model fire extinguisher can be made using the same chemicals.
   Have a student build one in class or at home. Some improvisation may be necessary and will provide modification for creative thinking. Have the student demonstrate to the class. The very common soda-acid extinguisher works on the same principle.

Evaluation

1. Can most youngsters give examples of physical and chemical change that occur in their environment?
2. Were the youngsters delighted to see the candle flame go out? Could many give a hypothesis explaining why it went out?
3. Do the students show some satisfaction knowing the formula CO_2?
4. Have you worked out a way to relight the candles quickly? Have you passed the idea along to other teachers?

Popping a Corn Seed

Specific Concepts/Skills

1. Heat helps change some things rapidly.
2. Corn seeds contain moisture.
3. Heat may cause corn seeds to pop.

MATERIALS

Per class
extra popcorn seeds
matches
popcorn maker (optional)
salt and butter (optional)
cooking oil—a drop placed in
   each test tube will help
   distribute heat and facilitate
   cleaning the test tube

Per team
1 small test tube
1 clothespin test-tube holder
1 2-cm × 2-cm (1″ × 1″) piece
   aluminum foil
1 7-cm × 7-cm (3″ × 3″) piece
   aluminum foil
1 candle
3 "popcorn" seeds
1 wooden stick (split tongue
   depressor)

VOCABULARY
gradual, rapid

The Activity

1. Say: "Today we are going to investigate the changes that occur in corn seeds when they are heated."
2. Pass out the student materials. Ask the youngsters to make a cap to cover the test tube by using the small piece of aluminum foil. Demonstrate how to do it. The cover will be used later.
3. Ask: "Are the corn seeds moist or dry?" (They appear to be quite dry.)
4. Have the youngsters put the clothespin around the test tube. A slight rotation of the tube will help get it into place.
5. Have youngsters light candles, drip a little wax on the center of the foil, and set candle in the melted wax.
6. Add one drop of cooking oil to each test tube.
7. Say: "Place one corn seed into the test tube and cover it with the aluminum foil. You are going to heat the test tube, but you must keep it about an inch above the flame."
8. Ask: "What will happen if the test tube touches the flame?" (Through previous investigations they learned the tube would be blackened and consequently they couldn't observe the change in the seed.)
9. Say: "When heating a test tube, *always point the mouth of the tube away from yourself and anyone nearby.* Now, place the test tube above the flame, and observe what happens to the corn seed and its surroundings. To keep the seed from burning, you might want to rock the test tube back and forth. (As the seed is heated, moisture will escape and turn to steam. The steam may then condense near the top of the test tube, and the seed will soon pop. When it does, the student should extinguish the flame and remove the popped corn when the tube is cooled (use the stick).
10. Ask:
    a. What changes have occurred in the corn seed? (Larger, different shape, softer, different color, etc.)
    b. Were the changes *gradual* or *rapid?* (Rapid.)
    c. What did you do to help make the change? (Added heat.)
    d. Why did the corn seed pop? (The seed contains moisture, as observed near the top of the tube. The moisture is enclosed in the tough outer cover of the seed. When the seed is heated, the moisture turns to steam. The steam exerts pressure on the cover. Usually the cover breaks at once and the corn pops open.)
    e. Why don't some seeds pop? (Outer cover is broken and no moisture within. Cover has cracks in it, and steam escapes without building up enough pressure to pop the seed.)

For More Mature Youngsters

f. Has the corn gained weight? Prove it. (No, it is larger but loses moisture and therefore weight.)

g. If a hole is poked into the corn seed, and then it is heated, what will happen? Try it. (It usually will steam and not pop.)

h. If we moisten the seeds for a day and then heat them, what changes do you predict will occur? Try it. (Usually pop sooner.)

11. It would be fun to end the lesson by making popcorn for the class and serving it with melted butter and salt.

Investigations at Home

Give each youngster a few popcorn seeds to take home. Under parent supervision they might like to demonstrate at home how and why popcorn pops.

Evaluation

1. Did most youngsters observe that the seed contained moisture?
2. Could most youngsters name at least four rapid changes that occurred when the seed popped?
3. Could one fourth of the class explain why the corn popped?
4. Do you notice a general improvement in laboratory techinques and conduct?
5. Will you long remember this lesson when you munch popcorn at the movies? Will it spoil the show or make you feel your efforts are worthwhile?

Heating a Raisin

Specific Concepts/Skills

1. Raisins contain water, sugar, and other substances.
2. Heat causes raisins to change.
3. Heat may be used to concentrate the sugar in a solution.
4. (optional) Excessive heat causes the sugar to change chemically.

MATERIALS

Per class
1 tsp granulated sugar
cooking oil (see previous lesson)

Per team
1 candle
1 small test tube and test-tube holder
1 7-cm × 7-cm (3″ × 3″) square aluminum foil to rest candle on
1 2-cm × 2-cm (1″ × 1″) square aluminum foil to cover test tube
several raisins
1 wood stick (split tongue depressor)
matches

VOCABULARY

concentrated, chemical change (optional)

The Activity

1. Say: "We are going to investigate the changes that occur in a raisin when it is heated."
2. Pass out the student materials. Ask the youngsters to make a small cap out of the piece of aluminum foil to cover the test tube. Demonstrate how to do it. The cover will be used later.
3. Say: "Chew one raisin slowly and try to remember the taste. You will compare the taste of this raisin to the one you heat."
4. Say: "Set your candle in the large square of aluminum foil and light it. Place a raisin into your test tube and heat the test tube over the candle flame. Remember, never point the test tube at any one nearby. Be sure that the test tube is held an inch or more above the flame to keep it from getting black. While heating, carefully observe what is happening to the raisin and to the inside of the test tube."
5. Caution the youngsters to heat the test tube slowly and to move it a little to distribute the heat.
6. Typical observations:

   The raisin has moved.
   I hear sounds.
   There's smoke (steam).
   There are water drops at the top of the test tube (condensed steam from the raisin).
   The raisin is getting bigger.
   The raisin is turning black.

7. Have the teams stop heating before the raisin gets black. A burned raisin is hard to clean out of a test tube.
8. After observing change, have youngsters blow out their candle. Place a clean piece of paper on their desk and tap the test tube against it. The raisin should drop out. Use the stick if necessary.
9. Say: "Use your senses to compare the heated raisin with an unheated raisin." (Smell, touch, and taste will reveal differences.)
10. Repeat the investigation, but this time place the aluminum foil cover over the test tube. (This helps trap the moisture, leaving the raisin.) Make careful observations.
11. After the candles are out and all the raisins eaten, conclude by having youngsters discuss the changes they observed, and the differences between the heated and unheated raisins. Ask such questions as:

    a. What is a raisin? (Dried grape.)
    b. Where did the steam come from? (Raisin contains water.)
    c. Why did the raisin move? (Steam escaping causes movement.)
    d. Some student thought the heated raisin tasted sweeter. How can you account for that? (Heat drove off some water, and the remaining sugar became more concentrated. Discuss meaning of *concentrated*.)

e. Some thought the heated raisin wasn't sweet at all. How can you account for that? (When sugar is heated to a high temperature, it changes and is no longer sugar. That kind of change is a chemical change. This may be demonstrated to the class by heating sugar in a test tube. As the sugar heats, its color changes, and it smells like caramel. Then it blackens and loses its sugar properties. It is difficult to clean the test tube and therefore unless test tubes are abundant, it is suggested only one be used.)

f. What differences did you observe with and without the aluminum cap? (Collection of water droplets is more obvious.)

g. If the class has done the popcorn investigation, ask: "Since both the popcorn and the raisin contained water, and both were heated, why didn't the raisin 'pop' like the popcorn?" (The outer cover of the raisin allows the water, as steam, to escape gradually.)

h. What is concentrated orange juice? How do you suppose it is made?

Investigations at Home

1. If grapes are available, find out how to change grapes into raisins. Try it.
2. With parental guidance, dissolve one tablespoon of sugar in one-half cup of water. Taste the water. Now find a way to make the sugar-water sweeter without putting anything in it. Explain to the class what you did, and why the water became sweeter. (Boil the water or let it evaporate. This concentrates the sugar and makes the sugar-water sweeter.)

Evaluation

1. Were most youngsters able to articulate the changes they observed?
2. The concept of concentration was introduced. Does at least one third of the class understand the meaning and significance?

MATERIALS

Per class
500 gm (1 lb) brown sugar
material for making pancakes
 (optional)
large pot for making syrup
(optional)
real or artificial flavoring—
 maple, vanilla (optional)
Per team
1 candle (regular or votive)
1 small test tube and test-tube
 holder
1 small piece of aluminum foil
1 7-cm × 7-cm (3″ × 3″)
 aluminum foil
1 plastic or wooden spoon
 (optional)

VOCABULARY

formula, relative (optional),
proportional (optional)

Investigating Formulas and Pancake Syrup

Specific Concepts/Skills

1. A formula is like a recipe in that it indicates the relative amounts of various substances required.
2. (optional) When the amount of one ingredient in a formula is changed, the amounts of the other ingredients must be changed proportionately.

The Activity

1. Say: "During this investigation you are going to change sugar crystals and water to pancake syrup."
2. Say: "We are going to use two different formulas to make the syrup. A *formula* may be thought of as a scientific recipe. It tells (relatively) how much of each substance to use." (You may wish to cite a simple formula to illustrate. The preparation of oatmeal is one they have used and is quite simple—2 cups of water, 1 cup oatmeal.)
3. Say: "The formulas we are going to use are (write on the board)":

Formula	Sugar	Water
A	1	1
B	2	1

   "This says formula A uses 1 part sugar and 1 part water. Formula B uses 2 parts sugar and 1 part water."
4. Ask: "In formula A, if you use one teaspoon of sugar, how much water should you use? (1 teaspoon) (Ask about formula B. If the class is mature enough, show that the formulas 1:1 and 2:1 are proportions, so that if you use 2 teaspoons of sugar in formula A you would use 2 teaspoons of water, etc. The *amount* of syrup would change but the proportion of ingredients would remain the same.)
5. Say: "Select the formula you would like to use. Write it down on a piece of paper. Fold the paper in half. One team at a time will come up with the paper and place the amount of sugar they need for its formula on to the paper. Then pour the sugar into your test tube." (Demonstrate. Some teachers prefer the teams to come one at a time; others will distribute test tubes and a supply of sugar to each team.)
5a. (optional) For more mature youngsters, vary the amount of sugar given for the formula they select, i.e., for formula B give them 3 teaspoons of sugar and have them determine the appropriate amount of water needed to keep the ingredients proportional. (1½ teaspoons.)
6. Say: "After adding the sugar to your test tube, add the amount of water your formula calls for."
7. Say: "Make an aluminum foil cap for the test tube. Poke a small hole in the top with your pencil point to let steam escape."
8. Say: "Cover your test tube with the cap. Light your candles (the usual setup) and heat the test tube. Remember, don't point the test tube at anyone. Be sure to keep it about 2 cm (1 inch) from the flame. When the liquid starts to boil, raise the tube even higher. Allow the liquid to boil for about one minute, then blow out the candle and allow the liquid to cool."

9. When the liquid is cooled (about 2 minutes), have the team members taste it. "Does it taste like pancake syrup?"
10. Ask: "What brand syrup does yours taste like?"
    "How could we give it a maple flavor?" (Add artificial maple flavoring.)
    "How could we give it a vanilla flavor?"
    "Which formula gives a sweeter syrup? Why?" (B, more concentrated sugar.)
    "How could we make formula A sweeter without adding sugar?" (Boil longer and drive off more water. The syrup will get thicker and sweeter as the sugar concentrates.)
    "Which syrup should cost less to buy?" (The less concentrated.)
11. (optional) Make pancakes or waffles (or ask cafeteria staff to do it, if you have unbelieveable rapport). Have each team add their syrup to their pancakes. You may wish to make larger quantities of syrup and then add maple or other flavor to it. Use the larger syrup for a class pancake party! The syrup also tastes good over ice cream or on bread.

Investigations at Home

1. Encourage the youngsters to experiment (under close parental supervision) making pancake syrup. Those that do can bring in some for class tasting. Have them write the formula on the board (unless it is a secret!).
2. Find out how pure maple syrup is made.

Evaluation

1. When given a new formula (recipe), can all youngsters understand it? If the quantity of one ingredient is changed (by a whole number amount), can the majority of youngsters adjust the amounts of the other ingredients proportionately?
2. Could the youngsters observe and describe the differences in syrup made by the two formulas?
3. Did most youngsters enjoy testing the product of their investigation?
4. Did any students report that they made syrup at home for use on Sunday morning pancakes?

Dissolving and Recovering Salt

MATERIALS

Per team
1 small test tube
1 hand lens
1 10 cm × 10 cm (4″ × 4″)
  piece black construction
  paper
1 eye dropper
1 candle
1 piece aluminum foil
1 small plastic cup
1 clothespin
salt

VOCABULARY
grain, evaporate, crystal

Specific Concepts/Skills

1. Salt grains have a cubic crystal shape.
2. Salt dissolves in water and may be recovered.

Note to Teacher

This is the first of three related lessons. Each lesson may be used independently. This lesson involves identifying salt by taste and crystal structure. Then the salt is dissolved in water and youngsters investigate ways of recovering the salt from the water. The next lesson investigates the effects of salt water upon plant life. The third lesson allows youngsters to attempt to solve the so far unsolved problem of separating salt from water and collecting the pure water in some economically feasible manner.

The Activity

1. Pass out materials. Have youngsters fill one plastic cup one half full with tap water.
2. Have the youngsters place a pinch of salt on the black (or dark) paper. (Binder paper is all right, but will not show crystals as well.)
3. Have the youngsters examine the salt with their hand lens. Have them draw what they see.
   Say: ''Do the salt grains look the same?'' (All should discover the many cubic-shaped crystals.)
4. Ask: ''Will the salt change if we add water to it? In what way(s) will it change?'' (Allow discussion.)
5. Say: ''Carefully pour the salt into your test tube. Use the eye dropper to add fresh water to the test tube. Each time you add water, shake the test tube. Continue until the water is clear.'' (Demonstrate. Youngsters can cover the mouth of the test tube with a finger or palm of the hand.)
6. Ask: ''Is the salt still in the water? How do you know?'' (They can't see it, but they can taste it. Let them.)
7. Ask: ''How has the salt changed? (The salt dissolved in the water. It has changed into such small pieces that it cannot be seen. But it is still salt.) (Optional: This kind of change is known as a *physical change*. The salt has changed its size but it is still salt. Can you think of other physical changes—tearing paper, chopping wood, etc?)
8. Ask: ''Can anyone think of a way to separate the salt from the water?'' (Allow discussion and if possible have them try their own methods. Often evaporation will be suggested and this investigation proceeds with that method.)
9. Say: ''Turn the dry plastic cup upside down and place *one drop* of the salt water on the cup.'' (Demonstrate using eye dropper.)
10. Ask: ''What do you think will happen to the salt water drop?'' (It will evaporate.) What will happen to the salt within the water?'' Set cup aside until step 16.
11. Say: ''It will take a while for the water drop to evaporate, so let's set it aside.''
12. Ask: ''Can you think of a way to make the water in your test tube evaporate quickly?'' (Heat it.)
13. Say: ''Heat the salt water in your test tube.'' (Demonstrate the usual method.

Caution youngsters not to point test tube at anyone and to hold it above the flame so as not to blacken it. *To cut heating time,* be certain test tubes are less than one-fourth full.)

14. After water has boiled away, have each team blow out the candle and wait for test tube to cool.
15. Ask: "What is in the test tube? How do you know?" (They may shake out the salt and examine with hand lens, taste it, or add fresh water to test tube and taste the water. Often the cubic crystal forms will not be observable.)
16. At this point, the drop of water on the back side of the cup may have evaporated. If so, have class examine the residue or set aside until later. It will be salt and usually in its crystal form. Tap the crystals onto the black paper and observe using hand lens. (If the youngsters have done about all they can and are getting restless conclude here with a summary. If not, continue to Part 2.)

Investigations at Home

1. Find out where the salt you use at home comes from.
2. Are table salt and ocean salt the same? (No, ocean salt contains our table salt, sodium chloride, along with other salts.)
3. Why is an iodine compound added to some boxes of salt? (People need very small amounts of iodine compounds in their diet. These compounds are found in fish and certain vegetables. Without sufficient iodine compounds, certain conditions such as goiter can develop. Iodized salt will prevent this problem.)
4. How can you keep salt grains in salt cellars from getting stuck together? (Add something to absorb moisture, such as uncooked rice.)
5. What is a cow lick? (Block of salt that cows lick.)

Evaluation

1. Were most youngsters able to observe the crystal form common to salt grains? Did they draw or describe the cubic shape accurately?
2. There was a lot to do in this lesson. Sometimes water is spilled, a youngster touches a hot test tube, or goes off on a tangent. Were you able to keep perspective, viewing the blunders as part of the growth process, and even enjoy yourself? (Then you're great.)
3. Are you taking notes that will help you or other teachers next time a class trys this lesson?

Investigating Using Salt Water on Land-growing Plants

Specific Concepts/Skills

1. All living things require water.
2. Salt water is harmful to many plants and animals living on land.
3. Some plants and animals require a salt-water environment.

MATERIALS

Per class

world map (optional)

2 carrots (potato and celery optional)

knife

1 MPC

salt

Per team

3 plastic cups

masking tape

3 ½ cm (¼'') slices of carrot (other vegetables optional)

For other materials see Investigations II, III, and IV

VOCABULARY

oceans (Pacific, Atlantic, etc.)

crops, nation, border

Note to Teacher

This is the second of three related lessons. Each lesson may be used independently. This lesson consists of four investigations. Investigation I should be done by the entire class. Investigations II, III, and IV are more appropriately conducted by volunteer teams. Investigations I requires observations to be made the next school day. Investigations II, III, and IV require observations to be made from one to ten days after they are begun.

The Activity

1. Say: "Many nations border on the oceans. (Demonstrate using map. Discuss oceans and other related topics appropriate at this time.) Yet, many of these nations don't have enough water to grow crops, to raise cattle, and at times, for people to drink. Why don't they use the ocean water to solve the water shortage?" (It may be necessary to point out that ocean water is salty. Often youngsters accept this, but think the real problem is moving the ocean water to the land. That technical problem can easily be solved using pumps. Guide the youngsters to realize that the real question is *what effect salt water will have on plant and animal life*.)
2. Ask: "How can we investigate the effect of salt water on plants?" (Allow youngsters to propose their own investigations. The following investigations are easily carried out by youngsters and may readily be modified to approximate their suggestions.)

Investigation I (to be done by the entire class)

Distribute team materials.

Have each team partially fill one plastic cup with salt water and one with tap water. Leave one cup empty.

Cut slices of carrot about one half centimeter (one fourth inch) thick. Slices of potato or celery may be substituted or used in addition. Give three pieces of each plant to each team. Have them place one slice of carrot in each cup. Use tape to identify each team's cups.

Bring out in discussion that the carrot (and other plants used) are living. They will be able to observe the effect of tap water, salt water, and no water upon the plants. Set the cups aside (in the sun if possible) for later observation. Have youngsters observe by sight and touch at least once before they leave school. (By the next day certain changes will be obvious. The plant slice in salt water may be discolored and soft. The plant slice in the tap water will be firm and have maintained its color.) Allow the youngsters to make their own observations and conclusions.

After the class has set up Investigation I, discuss the possibility of using entire plants and seeds to further investigate the effect of salt water upon plants. Investigations II, III, and IV may be conducted by volunteer teams. All materials needed are used in various other lessons and should be available.

Investigation II–Effects of salt water on germination of seeds*

Plant birdseed (or some other seeds) in cotton or paper toweling (see Lesson 2-1) in three separate containers, using any available containers. Moisten the cotton in one container with tap water; the cotton in another with salt water. In the third, use no water. Observable changes will occur within 3-5 days. (Only the fresh water seeds will grow.)

Investigation III–Effects of salt water upon growing plants*

Divide growing plants (such as birdseed, lima bean, pea, etc.) into three groups, in three separate containers. Water one container of plants with tap water, one with salt water, and don't water the remaining container. Observable changes will occur within 1-5 days. The plants watered with salt water will wilt and die. Those without water will wilt but can be revived.

*See Note to Teacher, page 98.

Investigation IV—Effects of salted soil upon plant growth*

Mix some potting soil with salt. Plant some fast-growing seeds (lima bean, birdseed, pea, radish, etc.) in salted soil, and the same kind of seeds in unsalted soil. Water each with tap water. Observable changes will occur within three to ten days. Once soil has been salted (as in parts of northern Mexico) it is useless for most crops until the salt is removed.

3. While some students are setting up these investigations, others may set about finding the answers to some of the following questions:

   a. What plants live in salt water? (Seaweed and all other sea plants.)
   b. What happens to the salmon as it swims from the saltwater ocean to a fresh-water stream? (Many films graphically show the changes.)
   c. What animals live in fresh water? What animals live in salt water? What might happen if they changed places? (Usually die.)
   d. Some plants and animals live in brackish water. What is that and where is it located? (Water found where rivers enter oceans. The fresh and salt waters mix.)
   e. What would happen to you if you went into salt water? Why? (Most people have and with little effect.) Their skin and fat below the skin protects them.
   f. At one time, all the oceans contained fresh water, and now they are salty. Where did the salt come from? (Streams dissolve salt from rock and carry it to the sea. This process is continuing and seas are becoming saltier.)
   g. Why do they spray water on vegetables and fruits in the markets? (Plants absorb the water and remain firm. Without the spray, they lose water and wilt.)

*See Note to Teacher, page 98.

h. Why doesn't salted meat spoil as fast as unsalted meat? The salt kills bacteria that would live on the meat and cause it to spoil.
i. Most people like crisp salads. How can you help keep lettuce crisp? (Spray with fresh water, shake off extra water, and refrigerate in covered container to prevent evaporation.)

Concluding Note

Each of these investigations contains elements of true experimentation. There is a problem to be solved (the effect of salt water upon seed germination). The youngsters have their hypothesis. They conduct a controlled experiment. That is, one container receives salt water while the other two do not. If the salt-watered seeds or plants didn't grow the experimenters couldn't be sure that the salt water prevented growth unless they had set up controls. They observe daily what occurs and finally make conclusions and a report to their peers.

Perhaps you might decide to take time to discuss the significance of the scientific processes they used in performing their experiments.

Investigations at Home

1. Ask the youngsters to repeat Investigation I, using any fruits or vegetables available at home.
2. Have a parent help with the following home investigation. (It's a good one for class use, but a bit expensive.)
   a. Prop two potatoes up in a glass of water using toothpicks. (Food coloring may be added to water if available.)
   b. Cut off the skin from the top and bottom.
   c. Scoop out about one-half inch of potato from the top of each.
   d. Fill the scooped out part of one potato with salt. Place nothing in the other.
      Place potatoes back into water. Observe what occurs. Report results to class.

Evaluation

1. Were all youngsters able to set up and evaluate the results of Investigation I?
2. Did the youngsters propose their own investigations? Were they worthwhile?
3. Was the room busy with "scientists" trying to solve problems through experimentation?
4. Were youngsters interested in learning the results of other team's experiments?
5. Did you help maintain their interest by being interested yourself?

Obtaining Pure Water from Salt Water

Specific Concepts/Skills

1. Salt and water can be separated and the desalted water collected.
2. It is often difficult to translate ideas into reality.

MATERIALS

Per team
(available to all)
1 candle
1 test tube
aluminum foil
eye dropper
2 plastic cups
2 test-tube holders
masking tape

(available on request)
1 gallon jar
drinking glass
dark sock or cloth
IMPC aluminum or glass tubin
Per class
salt water
distillation apparatus (borrowed
from junior high; optional)

VOCABULARY
irrigation, heat source,
evaporation, condensation,
distillation

Method 1

Note to Teacher

This is the third of three related lessons. Each lesson may be used independently. During this lesson youngsters will attempt to distill pure water from salt water. Immature youngsters may have difficulty solving the problem on their own, so the teacher might wish to select one method for the class to use and give materials necessary to perform that one investigation only.

Creative youngsters will relish the opportunity to solve the problem on their own and should be given the opportunity to try. The teacher may wish to set up one method (method 4 is a good one) and allow the youngsters to critique it.

The Activity

1. Say: "Salt water cannot be used to grow land crops. If the salt could be removed from the salt water, the remaining desalted water could be used to irrigate land that is now desert. If *you* can find a fast and inexpensive method of separating salt and water you will have solved one of the world's greatest problems. Your method can change the course of history and probably make you rich and famous!"
2. Ask: "Who recalls how we separated salt and water?" (Boiled off water; allowed water to evaporate.)
3. Ask: "What happened to the water during each process?" (Water went into the air.) "What was left?" (Salt.)
4. Say (or bring out in discussion): "Since the salt was left behind, the water that evaporated must have been desalted. If we could capture the evaporating water, it should be good to drink, free of salt, and therefore useful for crops. *But, how can we trap the evaporating water?*" (The problem.)
5. Say: "Your job is to find a way to get pure water from salt water. You may use any available materials, but first you must plan what you are going to do. Then, set up your equipment and finally, before you begin, have the teacher check your setup for safety."
6. Distribute equipment as needed.
7. Undoubtedly, some guidance will be needed for younger children. Discuss the need for a source of heat to speed evaporation. They have a candle and perhaps sunlight available. They also must remove heat from the evaporating water (steam) so it will condense as liquid. (See illustrations of student- and teacher-devised methods on pages 102 and 104-105.) You may wish to have younger children collectively use method 1 or 2, and later allow them to improve on those methods.
8. *Caution*: If youngsters make their own setup, be certain that the steam can escape from the collecting test tube or other container.
9. There probably will be many failures. *The important thing is for the youngsters to understand what they were trying to do and what went right and wrong.* In method

1 (the simplest and best for younger children), the steam may not condense on the foil and little water will be collected. *But*, have them taste the water on the foil or in the cup; it will be unsalted. Method 2 will involve similar problems. Method 3 works very well, except when the salt water boils up and enters the condensing tube. Then the salt will be carried by the steam into the collecting tube. Method 4 is slow and requires sunlight but it works every time. (Be sure you use a clean, dark-colored sock or absorbent cloth.)

10. After a sufficient period of investigation, have the teams explain to the class what they did and why. Encourage them to use terms such as: *source of heat energy, evaporation, steam, condense, equipment*, etc.

11. If distillation equipment is available, you may want to demonstrate it *after* the youngsters have worked out their own methods.

12. Ask: "Scientists (and class members, too) have solved the problem of separating salt from water and collecting the water. Why aren't these methods generally being used to obtain fresh water from the oceans?" (Too slow, too expensive, and uses too much fuel.)

Method 4

Method 2

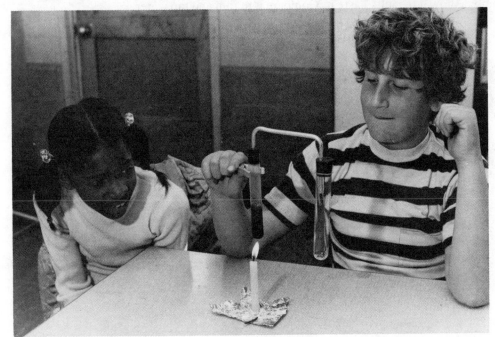

Method 3

13. Review with questions such as:
    a. Are clouds made of fresh or salt water? (Fresh.)
    b. Where does the water that form clouds come from? (Mostly evaporation of ocean water. The distilled ocean water returns to the land as desalted rain or snow.)
    c. Have you ever tasted water from a different place? How does it compare to your tap water? What do you suppose gives water its taste? (Dissolved minerals and perhaps city-added chemicals.)
    d. Where does our water come from?
    e. Some scientists have suggested icebergs be used as a source of fresh water. The icebergs would be towed to the place needing water and then melted. What do you think of that idea?
14. If possible, have the custodian take the class on a tour of the school to see the way water enters, is heated, distributed to rooms, and finally leaves the school.

Investigations at Home

1. Trap some fresh rain water in a clean container. Taste it. How does it compare with your tap water?
2. Trap some fresh snow. Allow it to melt. Taste it. Compare it to rain and tap water.
3. Ask the youngsters to have their parents show them the pipes that deliver water to their house and the pipes that carry away dirty water.

Evaluation

1. Did the youngsters show evidence of creativity?
2. When a team reported to the class, were their ideas respected by themselves and others?
3. Do you have the feeling the youngsters really understood that they were working on a significant problem which remains unsolved? Do you feel that way, too?

Suggested Level: ■ *K-2*　■ *3-4*　■ *5-7*

Approximate Time: Introduction – ½ hr
Setting up – 1 hr
Observation – 3 days to 3 weeks

Setting Up Student Aquariums

Specific Concepts/Skills

1. Living things require a suitable environment in order to live.
2. The plants and animals in an aquarium interact with each other.

> ### MATERIALS
> **Per class**
> 1 fish net (small)
> **Per team**
> 1 MPC
> 1 or 2 fish
> 1 or 2 sprigs Elodea (or substitute)
> pure sand
> water snail (optional)
> various decorations (optional)

> ### VOCABULARY
> aquarium, oxygen, carbon dioxide, fry, chlorine, bacteria

Note to Teacher

The beauty of this lesson is that each team of two youngsters can have their own aquarium for far less than the cost of one display tank. And when the in-class investigations are over, the fish and plants can be given to the youngsters to take home. Using this low-cost setup, investigations of various sophistication can be conducted. The investigations that follow are intended for younger students.

The Low-Cost Aquarium

1 multipurpose plastic container	50 cents
2 goldfish	30 cents
1 sprig Elodea	3 cents
Cost per complete aquarium	83 cents
Expendable cost	33 cents
Total expendable cost for 15 aquariums	$4.95

The Activity

1. Prepare the children well in advance for this lesson. Tell them they will be getting their own fish but must prepare the proper environment for them. Have the class decide what they must do to be ready for the fish.

2. List the required items as mentioned (aquarium, clean water, and plant). Some children may want to add sand, snails, a toy, fish food, etc. Distinguish between what is necessary and what is extra. In this aquarium, packaged fish food is *not* needed.

3. Discuss the function of each thing they will place in their aquarium, and what may occur in the aquarium. Some likely questions and their answers follow:

   a. Why does the aquarium need plants? (The plants are food for the fish, the plants give off oxygen gas needed by the fish, the plants take in carbon dioxide gas that the fish breathes out into the water and uses it to make its own food.)

   b. Why does the aquarium need clean water? (Polluted water may contain chemicals or germs that can harm the fish.)

   c. Where can we get clean water? (Tap water seems clean but a chemical called chlorine has been added to the water to kill any small germs [bacteria] that might live in it. The chlorine can harm the fish. We can rid the water of the chlorine gas by allowing it to stand 24 hours.)

   d. Why do fish need water? (They can move freely only in water; they can't breathe air the way we do—they must get their air [oxygen] from the water; they can only eat in water; they lay eggs [have young] only in water [goldfish won't reproduce in small aquariums; guppies will]).

e. Do we need sand? (Sand makes the tank look nice but it is not necessary. If sand is to be added [necessary for guppies], the following method has proven successful: Use only pure sand. Have youngsters place about 1 inch of sand at the bottom of a dry tank. Cover the sand with a paper towel and then slowly add water. Carefully remove towel so as not to disturb the sand.)

f. Shall we add fish food? (You can add food to a tank equipped with filters that will clean the water. In our small aquarium, however, any uneaten food will settle to the bottom of the tank. This food may allow small germs [bacteria and fungi] to multiply in the tank, and harm the fish. So, the only food the fish should get is the Elodea.)

g. Can we add just a little tiny bit of food? (Only if you are absolutely sure the fish eats it all.)

h. What if the water gets dirty, cloudy, or begins to smell? (These are all signs that the water is polluted, probably with bacteria or fungi. The water must be changed. Remove the fish and place in clean, chlorine-free water of about the same temperature water and return the fish.)

i. What if the water turns green? (Good. That means small green plants called algae are probably growing in the water. Fish use the algae as food. The green color won't bother the fish. If it bothers you, you may clean the tank.)

The Day the Fish Arrive

Be sure each team's tank is ready and labeled with their names. Contact the aquarium store and let them know when you will arrive. Try to arrange to pick them up the day you are going to distribute them. It is a good idea to buy a few extra fish to cover fatalities in transit.

Distribute the fish, one or two to each student aquarium. If the youngsters have flat-top desks, you may want the aquarium to remain on the desk all during the week(s) of investigations. It takes several days for the youngster and the fish to get acquainted, so don't begin the following investigations for two or three days.

Suggested Student Activities

1. Draw the fish and the things that make up their environment.
2. Draw a large picture of a fish and label its parts. (Have youngsters use a book to help identify the parts.)
3. Does the fish sleep? (Yes, in its own way.)
4. Does the fish blink its eyes? (No.)
5. How does it move? (Fins.)
6. Can it move backwards? (Yes.)
7. How many fins does it have? (Depends on kind of fish.)

8. Does the fish see in front of it? (Eyes are on side of head; may not see directly in front.)
9. Do you think fish get bored?
10. Do fish breathe? (Yes.) How? (Water is taken in mouth and passes out of the gill slits. Some oxygen is removed from the water in the gills and enters the fish's blood.)
11. Can you tell if your fish is a male or female? (Not with small goldfish.)
12. Do the fish seem to play?
13. Do the fish eat the plants? (They will if they are not fed.)
14. Have the plants grown? (They may grow more leaves and roots.)
15. Can you tell if the plants are giving off gas? (Yes. Bubbles can be seen.)
16. Can you tell one fish from another? What makes them look different? Do they behave differently?

Suggestions for Use of Guppies

If guppies are available, they may be substituted for goldfish. They are smaller and require more care, but they bear live young and that can be fascinating. Guppies are warm-water fish and require temperatures above 21°C (70°). If the classroom is colder than that on weekends, it would be unwise to use guppies without providing a source of heat. Commercial heaters are expensive and would skyrocket the cost of the lesson. However, some teachers place all aquariums together (away from windows) and heat with two or three lightbulbs placed above the tanks. It is important to experiment with temperature control before you get the guppies. In addition, each tank should have additional plants and pure sand on the bottom.

Additional Investigations Related to Guppies

1. Each tank contains a male and a female; can you tell which is which? (Male more beautifully colored; female has larger abdomen.)
2. How can you tell if the female is going to have *fry* (baby fish)? (Female's abdomen gets greatly enlarged and a black spot may appear.)
3. Are the fry released as live young or eggs? (Live young.)
4. Do the male and female care for the young? (No, they may eat the young.)
5. How do the young protect themselves? (Hide among the plants.)
6. How many fry did the female have? How many remained after 1 day, 2 days, etc.?
7. How can you help protect the fry? (Transfer to another tank or feed the parents enough food so they will not go after the young. Caution must be taken not to overfeed and pollute the water.)

Investigations at Home

At the conclusion of the in-class investigations, the fish and plants may be sent home with the youngsters. Clean mayonnaise or mustard jars, etc., can be used as tanks.

Evaluation

1. Did most youngsters understand the need for a suitable environment?
2. Were most youngsters able to observe and understand the interactions between the plants and fish?
3. Did youngsters improve their ability to observe, as indicated by their responses to observation investigations?
4. When a fish died, how did the owner respond? How did the class respond? What did you say?
5. Was your work worth the effort?
6. Were the aquariums on display for open house? How did the parents respond?

Studying Earthworms

Specific Concepts/Skills

1. Earthworms are living things, all living things are unique.
2. Earthworms have common and individual characteristics.
3. Earthworms sense their environment and make choices. These choices are usually predictable and in all probability are inborn.
4. Earthworms are useful to man.

MATERIALS

Per class
paper towels
small garden tools (optional)
soil, peat moss, sawdust
scalpel or razor
Per student or team
1 earthworm per student
2 MPC covers per team or 2 complete MPC's
1 hand lens
1 magnet

VOCABULARY

(all optional) bedding material, blood vessel, clitellum, warm blooded, cold blooded, living, nonliving, egg case, dorsal, ventral, anterior, posterior, anus, segments

---

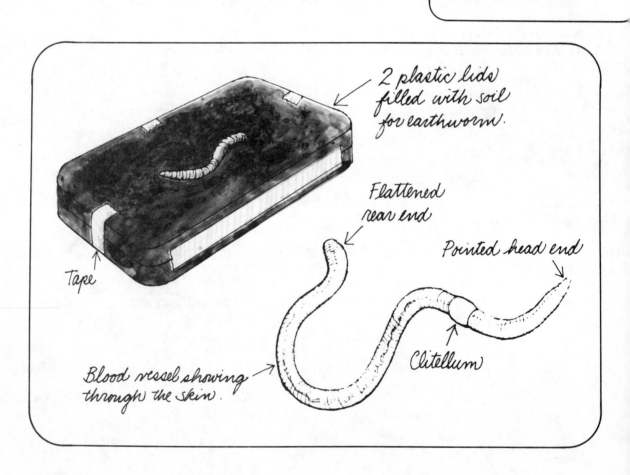

2 plastic lids filled with soil for earthworm.

Tape

Flattened rear end

Pointed head end

Clitellum

Blood vessel showing through the skin.

Note to Teacher

Lessons 2-16 and 2-17 involve the study of small living things over a relatively short period of time. The two lessons should not be used consecutively, but, rather, between other units. There are various ways of handling these investigations, ranging from the unstructured ("Here they are, find out what you can") to the super-structured ("Your lumbricus has been preserved for you in formaldehyde. The worksheet will tell you exactly what you are to do and what you will discover. Report your discoveries in the appropriate discovery box.").

Our experience indicates that no strategy is best for every teacher. Even the right strategy for one class will change as the dynamics within that class change. This is particularly true when investigating living things. Therefore, the investigations in this unit are written so as to provide guidance to the teacher, without the structure of previous units. The guidance involves four areas:

1. Low-cost materials are recommended that have been used in previous investigations. This minimizes cost and effort, making it possible for every youngster to investigate the organism.
2. Investigations are those that youngsters can realistically be expected to conduct and methods are given that a teacher may use to prompt student interest.
3. Facts and concepts are outlined that youngsters should encounter and, to varying degrees, assimilate.
4. Attitudes that youngsters should develop when working with living things are suggested.

The recommendations in items 1-3 have been extensively classroom tested and are a synthesis of ideas that work and concepts that are generally assimilated. Recommendation 4 is the author's bias, sometimes rejected by others but presented without compromise.

The Activity

1. Say: "Today we are going to begin the study of something very special. Let's see if you can guess what it is, as I give hints." (Some teachers have youngsters write guesses down after each hint.)

One   — It uses fuel to move, but it does not pollute the air.

Two   — It contains a guidance system that allows it to choose direction based upon its environment.

Three — It's very flexible.

Four   — You could spend billions of dollars, get the help of the best scientists and equipment and never build one; but each of you will get one today.

Five   — It's both male and female at the same time.

Six   — It is useful to birds, gardeners, and fishermen.

Seven — Although man can never make one, an earthworm can.

2. Discuss the need to prepare a home in class for the earthworm. Ask: "What do you think the earthworm needs?" (The answer can come from discussion or a brief field trip to find earthworms and to note the environment in which they were found.)

3. Containers for worms and bedding material (soil, peat moss, or experimental material) will be needed. Two recommended containers are:

   a. Place a thin layer of bedding material in the cover of one MPC. Use masking tape to hold a second cover on top of the first. The tape should be placed along one long side. This will act as a hinge. Use smaller pieces of tape for the other sides (see illustration). This method will allow for observation without having to dig the worm out of the bedding material.

   b. Place soil or experimental material in an MPC and cover. This is the easiest and cleanest method, but worms will have to be removed for observation unless a very thin layer of bedding is used.

4. Either have youngsters go outside to find earthworms or distribute the purchased worms.

5. It is not unusual for some youngsters (or teachers) to be reluctant to handle earthworms. Allow them to observe others at work and invariably they will soon join in.

Recommended Investigations

I. Getting to Know You

   A. Class is arranged in teams of four. Each youngster has his own worm. Worms should be removed from the home and placed on MPC cover or paper towel for study. The following questions are designed to encourage careful observation.

      1. Compare your worm with your team members' worms. How is yours like theirs? What makes yours different? Be sure you can tell them apart because they will be sharing the same home.

      7. Study your worm. How can you tell its front (head end) from its rear? Its top from its bottom? (Top is darker than bottom and worm moves bottom side down.)

      3. Use your hand lens and try to find out if your worm has:
         a. Eyes. (No.)
         b. Mouth. (Yes.)
         c. Clitellum (Band circling the body near its front end; useful to youngster in determining front from rear; useful to worm for reproduction.)
         d. A blood vessel running the full length of the worm? (Yes.) What color is it? (Purple.) Do you have blood vessels? Can you see any of yours? (Wrist.)

e. Is the worm smooth? (It's divided into segments.)

f. Does it have legs? (Very difficult to see, but leglike structures can be found on the underside near head of worm.)

g. How does it move?

h. What does it smell like?

i. Is it attracted to a magnet?

j. How does it feel to you?

II. Preparing for the Big Race

A. Say: "We are going to find the fastest worm in the room. Each of you can enter your worm in the contest. The winning worm and his owner will be awarded a delicious treat. The worm will be given what he likes to eat best and the owner will receive the same. To help your worm, you should find out what will make him move fast and straight. Allow discovery on their own or suggest the following:

1. Does it move faster when warm or cold? (Worms are cold blooded. Their body temperature changes with environment. A warm worm is apt to move faster. Worms can be warmed or cooled by placing in water of appropriate temperature for about 30 seconds.)

2. Does touching it make it go faster?

3. Can you make a path of some kind that it will be apt to follow?

4. Does it prefer moving near the edge of the plastic cover or near the center, or doesn't it have a preference?

5. Does it go faster moving uphill or downhill?

B. Have team members race their worms and select a team winner to enter the Big Race.

C. Have team winners compete for class championship. Be sure "ground" rules (a pun) are established. Allow youngsters to apply what they learned in their investigations, before and during the race.

III. Is it a Male, Female, or What?

A. Youngsters may have discovered worm egg cases either in the soil outside, or later, in the worm's home. In either "case" (another pun), use the appearance of egg cases or new young as impetus to discuss reproduction of earthworms. Use reports, charts, and teacher-led discussion. Children find it fascinating to learn that their worms are both male and female, and the way in which they mate.

IV. Experimenting with Worms

A. Investigation can be conducted to find answers to the following questions:
1. Do worms prefer wet or dry soil? (Wet.)
2. Do worms prefer darkness or light? (Darkness.)
3. Why do worms come to the top of the soil after a rain? (They need air and soil is saturated and air is not available.)

V. What Do We Do with Them after We're Finished with Them?

A. Reemphasize the significance of *living* things. Discuss and review what has been learned.
B. Discuss what can be done with the worms. Many may wish to return the worm to its natural environment. A brief field trip may be taken, during which each youngster finds a suitable environment for his worm.

Investigations at Home

1. One or two students might be interested in raising earthworms. At the conclusion of these investigations give all the earthworms to them to start their enterprise.
2. Research project: Find the many ways earthworms are useful to man.

Evaluation

1. Did all youngsters get used to the worms and handle them with maturity? Did you?
2. Were youngsters able to identify their own worms?
3. Do most youngsters indicate through word and deed a respect for the earthworm as a complex living thing?
4. When the investigations were finished, did most youngsters show some concern for what would be done with the earthworms?

Studying Mealworms for Fun and Profit

Specific Concepts/Skills

1. The mealworm passes through four distinct stages in its life cycle—egg, larva, pupa, and adult.
2. Mealworms, like all living things, require a proper environment in which to live.
3. Mealworms may be investigated using controlled experiments (optional).

MATERIALS

Per class (all optional)
refrigerator
extra mealworms
beetles
dark paper
protractors
thermometers
rulers
graph paper
photos of people of various
 ages
Per student
plastic cup or substitute
clear plastic wrap (to cover
 cup)
2 mealworms
toothpick
2 tbsp oatmeal
small piece apple or potato
hand lens
MPC cover

VOCABULARY

life cycle, adult, egg, larvae,
 pupa, control (optional)

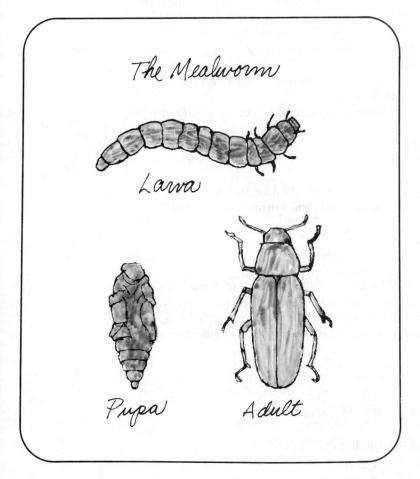

The Mealworm

Larva

Pupa   Adult

Note to Teacher

Mealworms are excellent classroom animals because they are readily available, inexpensive, hardy, easy to handle, and should go through a complete life cycle within a month. They may be purchased at local pet shops or in quantity from breeding houses. They can be housed in almost any container that permits some air circulation and keeps light out. Mealworms eat oatmeal and require a piece of apple or potato for moisture.

Introducing the Activity

1. Ask: "Which came first, the chicken or the egg?" (An old philosophical question for grownups, but new, provocative, impenetrable for many youngsters. During the discussion, bring out the concept of life cycle, from egg to chicken to egg.)
2. Ask: "What kind of life cycle do humans go through?" (Use photos from magazines to illustrate the changes that occur with increasing age.)
3. Say: "We are going to study a very interesting animal. We will find out as much as we can about its life and its life cycle. It is called a mealworm. The first thing we must do is provide a proper environment for the animal."
4. Pass out materials. (Container, cover [plastic wrap], oatmeal, piece of potato or apple, toothpick, rubberband.)
5. Ask: "What do you think each of the materials is used for?" (Apple or potato provides moisture; toothpick is used to punch holes in the plastic cover to allow air circulation.)
6. After environment is prepared, provide youngsters with two mealworms and a hand lens. If teams share containers, fewer materials and worms will be needed.

Recommended Investigations

I. Getting to Know You

A. Place a mealworm on a paper or MPC cover. Draw a picture of it and label its parts. (Teacher can write on the board words like *head, legs, segments,* etc.)
B. Allow youngsters to spend time discovering what they can about the mealworm, using the hand lens. You may wish to direct their attention with questions such as:
   1. Can it move backwards?
   2. What does it smell like?
   3. How does it feel to you?
   4. Does it sense its environment? How do you know?
   5. How does it eat?
   6. Can you tell one mealworm from another?
   7. Does it have a mouth, eyes, etc.?

II. Experimenting with Mealworms

A. Discuss things that can be discovered about mealworms by investigation. Guidance will be needed in selecting problems that can be solved and in setting up experiments to "solve" them. Allow the youngsters as much freedom as possible. You might suggest they find ways to answer the following questions:

a. Do mealworms prefer light or dark (see illustration)?

b. Some animals can move upside down (a housefly); others find it difficult to walk a steep incline (man). How steep an incline can a mealworm move up? Does it matter what the incline is made of (see illustration)?

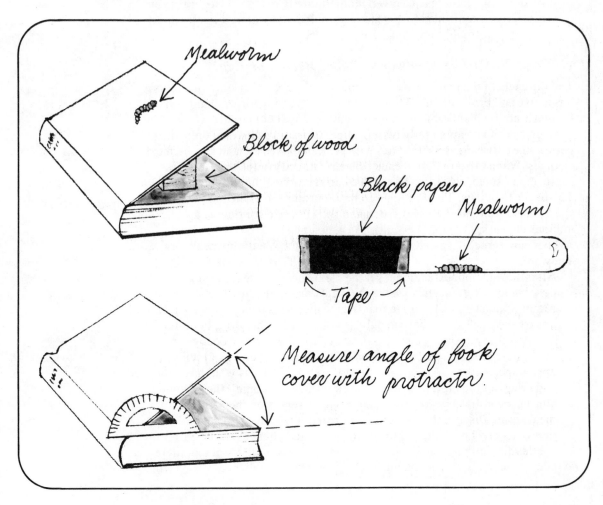

c. Do mealworms recognize color?

d. Can mealworms detect water? Will they enter water? Does the color of the water, or what is in the water, affect the mealworm's reaction?

e. Do mealworms grow? How much? How does temperature affect the mealworm's activity? (Mealworms are cold blooded. When their environment is hot 25°C [77+ F], their body temperature is also 25°C and they are very active. If cooled by refrigeration or by placing their container upon ice, they will be very inactive. Guide more mature youngsters in developing a quantitative measure of activity, e.g., distance traveled [in any direction] for 5 minutes at a given temperature. Continue tests at varying temperatures controlled by masked lightbulbs, hot water, ice, etc. and compare [by graph] temperature *vs.* activity.)

III. *Which Came First: The Mealworm, Pupa, Beetle, or Egg?*

A. The life cycle of the mealworm is absolutely fantastic. As a larva (mealworm), it is active and busily eating. Then, it slows down and begins a change that is far more intense and dramatic than the famous Dr. Jeckyl to Mr. Hyde. The larva changes to an apparently resting pupa. But during this seemingly inactive period, the entire organism seethes with profound physical and biochemical changes. When the pupa stage ends a black, fat, and completely harmless beetle (adult stage) emerges. The beetle can reproduce, laying eggs in the meal. The beetle then dies. The tiny eggs contain the nucleus of life and the plan for species perpetuation. The eggs hatch and the life cycle continues, as it has for millions of years.

There are three suggested ways to study the life cycle of the mealworm. They are:

1. After the initial investigations, set the mealworms aside for casual observation. Within a few weeks the mealworms will enter the pupa stage. Many youngsters will assume that the mealworm is dead. However, careful observation will show that the pupa is not a dead mealworm, but rather, a mildly active physically changed "mealworm." When the beetles appear, youngsters are surprised to see them, but often don't connect their appearance with the disappearance of the pupa. Discuss what has happened and bring out the life cycle that is unfolding in the "home" they provided.

2. During the initial investigations, have a team report on the life cycle of a mealworm. Discuss the changes anticipated. Over the next several weeks, youngsters will be delighted to see the predicted changes occur. After the beetles die, nothing much seems to happen. But remind them that there is a good chance eggs are in the meal. Hopefully very small mealworms will appear.

3. After the initial investigations with mealworms, pass out the beetles, without explaining the connection between the two.* Have them provide the same kind of environment for the beetles (or have only one container for all beetles as doing this increases chances of producing eggs). Youngsters can compare and contrast beetles and mealworms. Then use method 1 or 2 above to have them discover the relationship that exists.

Investigations at Home

1. Mealworms are used as food for many small laboratory reptiles, amphibians, and birds and there is often a local market for them. Perhaps a youngster or two would be interested in raising them for profit. They will learn a lot and earn some money with a very small investment.
2. Have the youngster find out about life cycles of such animals as: mosquitoes, frogs, butterflies, flies, kangaroos, earthworms, grasshoppers. Post the reports and illustrations on the bulletin board.

Evaluation

1. Do the youngsters indicate an understanding of the mealworm's life cycle?
2. Did the youngsters appear eager to learn the names of the four stages in the life cycle, or did you feel you were pushing the terminology? Did you keep pushing? Why?

*It is sometimes difficult to buy the beetles. One way to assure a supply is to obtain mealworms several weeks in advance. Keep some mealworms refrigerated in a closed container with meal and an apple or potato. Their life cycle will be interrupted and they will remain in the mealworm stage. Keep other mealworms at room temperature, allowing them to progress to the pupa stage. Then begin the mealworm investigation with the refrigerated mealworms. Soon the pupae will change to beetles. They can be studied simultaneously with the mealworms.

THEME **3**

Energy, Fields, and Forces

The concepts of energy, fields, and forces can be extraordinarily difficult. These concepts can seem nebulous, as they are difficult to define, touch, or measure. They do, however, affect us in every way and at all times.

The investigations included under this heading are designed to increase the youngsters' awareness of the effects gravity, magnetism, electricity, sound, and light have upon them and their environment.

Gravity has become a household word frequently associated with space and space travel. The selected investigations bring gravity home to earth by alerting the youngsters to its effects upon plants, animals, and themselves. Ways to overcome gravity are investigated and youngsters are guided in "discovering" static electricity and magnetism. They then use these forces to temporarily overcome the force of gravity.

Subsequent investigations involve the fields and forces of electricity, magnetism, sound, and light. Youngsters apply what they have learned in building a compass, flashlight, string phone, musical instruments, an electromagnet, and a remarkable motor.

These investigations are apt to be so enjoyable that you run the risk of having the youngsters miss the point. If this be the case, the teacher must gently redirect attention back to the elusive concepts of energy, fields, and forces.

MATERIALS
Per class
2 clear plastic cups

VOCABULARY
force, gravity

Investigating Gravity and How to Overcome It

Specific Concepts/Skills

1. A force is a push or a pull.
2. Gravity is the force that causes things to fall to earth.
3. Other forces can work against gravity.
4. Plants and animals can work against the force of gravity.

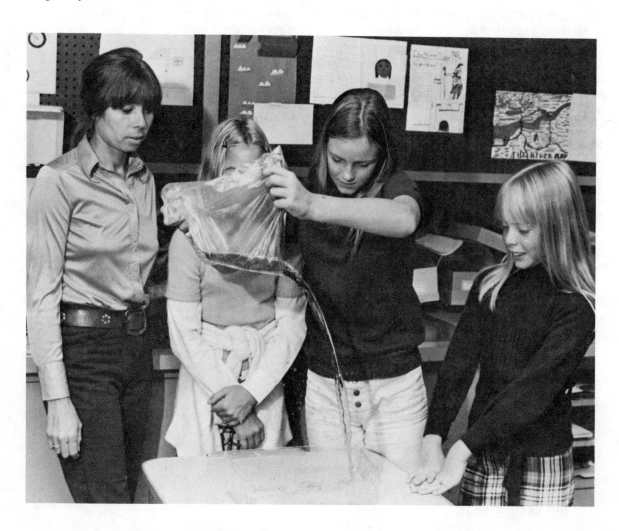

Note to Teacher

Students are introduced to the concept of gravitational force by observing and recalling its effects upon them and things in their environment. Then in this and subsequent investigations youngsters will attempt to overcome gravity using the force of moving air, static electricity, and magnetism.

The Activity

1. Ask one child to come to the front of the class and pour water from one glass to another.
   Ask: "What happens to the water?"
        "Why does it go down?"
        "Why doesn't it fall up?" (Let children give their own answers.)
2. Have several children bring up an object from their desk (pencil, eraser, crayon, etc.). Ask them to hold it up and then let it go.
   Ask: "Why did the things fall?"
        "Why didn't they go up?"
   Explain that a force is a push or a pull. Gravity is a force that cannot be seen, but it causes everything to fall down or be pulled toward the earth.
3. Have the children try to overcome the force of gravity by jumping, etc.
   Ask: "Can any of you overcome gravity?" Explain that when they jump, their muscles help them to temporarily overcome gravity. But then gravity "wins" and they come back down.

4. Ask: "Can any of you think of an animal that can overcome the force of gravity?" (Youngsters usually think of birds and stop at that. Encourage them to think of other animals.)
    Ask: "When does an ant overcome the force of gravity? (When it walks uphill.)
        "When does a kangaroo overcome the force of gravity?" (When it hops.)
    The children should soon realize that all animals and plants overcome the force of gravity to some extent.
5. Ask: "When you jump off a diving board why don't you fall up?" (Gravity.)
        "How does gravity help you when you hike?" (When you walk downhill gravity helps you move easily.)
        "How does gravity slow you when you hike?" (When you walk you must work against the pull of gravity. Walking uphill requires even more effort.)
        "If you live on the second floor of an apartment building, when does gravity help? When must you overcome it?" (Walking downstairs; walking upstairs.)
6. Pass out drawing paper. Ask the youngsters to draw a picture of an animal overcoming gravity. Have them also draw a picture of an animal in a situation in which gravity is a helpful force.

Evaluation

1. Did youngsters participate in the discussion?
2. Do they correctly use the words *gravity* and *force*?
3. Did they enjoy themselves?
4. Did their drawings indicate an understanding of how gravity is helpful to animals and how it is overcome?

Using Air and Static Electricity to Overcome Gravity

MATERIALS

Per class
1 large teacher balloon
5 cm (2″) plastic tube
1 MPC
1 Styrofoam cup broken into
   very small pieces
1 ball of string
small objects such as washers
   or paper clips, to be used as
   weights
Per student
2 easily inflated balloons
1 hand lens (optional)

VOCABULARY
air, static electricity

Specific Concepts/Skills

1. Moving air can overcome the force of gravity.
2. Static electricity can overcome the force of gravity.

Note to Teacher

This lesson continues the study of gravity and introduces the forces of moving air and static electricity. Students discover how these forces can temporarily overcome the force of gravity.

The Activity

1. Pass out an uninflated balloon to each youngster.
2. Say: "Scientists must think and work to solve problems. I have a problem I want each of you to work on and solve as best you can. I would like to find the best way to help the balloon overcome the force of gravity."

3. Allow the youngsters time to experiment. Then have them come to the front of the room and demonstrate their methods. Some methods include:

   a. Inflating and tying the balloon with string. (It falls more slowly because the air in the room helps keep it up.)
   b. Inflating and letting the balloon go. The balloon moves erratically but quickly falls down.

   At this point you may wish to demonstrate the same effect using the more stable "teacher balloon" with plastic weight (see illustration).

   c. If the youngsters have not yet discovered static electricity you may guide them as follows: Have each student inflate and tie a balloon. Then tell them to rub their balloon against their hair or sweater. (This will charge the balloon with static electricity.) Then have them place the balloon against a wall, their desk, themselves, etc. (It will stick.)
   Say: "When you rubbed the balloon, you helped make another force. It is called static electricity."
   Ask: "Can static electricity help the balloon overcome gravity?" (Yes.)
   Ask: "Can you use the balloon to help other things overcome gravity?"
   Pass out *very small* pieces of a Styrofoam cup. Have the students rub the balloons and then have them make the Styrofoam overcome gravity. Allow the youngsters to take the charged balloons out of the school yard. Have them move the balloon just around the ground. Many small things will jump up and stick to the balloon. Ask them to try and identify the small pieces.

   Allow the youngsters to take their balloon home. Tell them to charge their balloon and then find out what it will lift against the pull of gravity. Ask them to make a list of all things the charged balloon will lift. (Salt, paper, dust, hair, etc.) Incidentally, when a charged balloon is placed just above some sprinkled salt, the salt jumps to the balloon, making a beautiful sound.

Evaluation

1. It's difficult to keep calm with 30 balloons flying around the room. Were you able to control the situation?
2. Did the youngsters have fun with the balloons? Did they realize that they were using moving air to overcome gravity?
3. Do the youngsters use the terms *static electricity, gravity,* and *force* in their own conversations?
4. Are some youngsters after you to find out more about static electricity? Have you directed them to a useful source of information?

MATERIALS
Per class
helium gas
carbon dioxide gas
3 balloons
3 pieces string approximately
  180 cm (6 ft) long

VOCABULARY
predict

Showing How Helium and Carbon Dioxide Balloons Behave Differently

This is a good demonstration but it requires gases not normally found in elementary schools. Perhaps a kindly secondary teacher will lend you the gas containers.

What to Do

1. Fill three identical balloons to approximately the same size with three different gases such as helium, carbon dioxide, and air. Tie a 180-cm string to each.
2. Have three youngsters stand on their desks. Have each one hold a balloon in one hand and the corresponding end of the string in the other. Do not tell the class which gas is in each balloon.
3. Ask the class to predict what will happen to the balloons when they are released. Encourage them to use the term *gravity*.
4. Have the three youngsters release the balloons at the same time.
5. Ask the class to describe what happened and why.
6. The helium-filled balloon is lighter than air and floats in air as a piece of wood would in water. The carbon dioxide-filled balloon is heavier than air and sinks quickly. The air-filled balloon sinks slowly.
7. Ask: "Does the helium balloon overcome gravity?" (It appears to; it rises because the balloon and helium is lighter than an equal volume of room air. This is difficult for youngsters to understand. Perhaps it is best just to say it is lighter than air and floats.)

Air filled balloon sinks slowly.

Helium lighter than air. Balloon floats.

Carbon Dioxide is heavier than air Balloon sinks.

Making a Pin Float in Air

MATERIALS
Per class
1 spool of thread
several brass-filled pins
Per student
1 student magnet (see Appendix)
2 or 3 iron straight pins

VOCABULARY
magnet

Specific Concepts/Skills

1. The force of a magnet can work against the force
   of gravity.

Note to Teacher

This investigation continues the study of gravity and introduces magnetism. The force of a small magnet is used to make a pin float in air, thereby temporarily overcoming gravity.

The Activity

1. Say: "We have studied gravity on earth and how it pulls all things down to earth."
2. Ask: "What things have we studied that can overcome the force of gravity?"
   (Plants and animals, moving air, static electricity, and helium-filled balloons in air.)
3. Say: "Today we will study something else that can overcome the force of gravity." Hold up one student magnet. (Don't tell the children what it is. It does not look like the usual magnet.)

4. Ask: "How many know what this is?" Pass out one magnet to each youngster. Allow them to feel it and examine it closely. If two magnets are brought together, their properties will be obvious. But let the children discover this!
5. Pass out two or three of the iron pins to each child. Allow them time to discover the properties of the magnet and pins.

   Occasionally, you may have a student that requires an additional challenge. Try this: Obtain three brass-filled pins (the most common type). When passing out pins to the other youngsters, hand the three identical looking but nonmagnetic brass pins to the students needing additional challenge. At first, they may believe their magnet is different, but encourage them to investigate further.
6. Ask: "Does the force of the magnet cause the pin to overcome the pull of gravity?" (Yes, it jumps up towards the magnet.)
7. Tell the children that you will show them how to make a pin float in air. Demonstrate and then have the youngsters try it:

   a. Tie one end of a 15-20 cm (6-8 in) thread to a pin.
   b. Tie the other end around a pencil, crayon, or any other available object.
   c. Stick the magnet to metal, such as the desk legs, and touch the pin to it. The pencil should rest on the floor.
   d. Slowly pull the string until the pin pulls away from the magnet and "floats" freely in air (see illustration).

Investigations at Home

1. Allow each child to take home his magnet, pins, and string. Encourage students to show their parents how to make a pin "float."
2. Ask them to make a list (or drawing) of each thing their magnet will stick to at home.

Evaluation

1. Can the youngsters name several different ways to overcome gravity?
2. Do they realize that gravity pulls on *all* things, while magnets only pull on certain metals? (Iron, steel, and a few others.)
3. Were they able to "float" their pin?
4. Did they enjoy themselves? Did you enjoy yourself?

Fishing for the Truth

MATERIALS
1 pole*—1-2 meters (3-6 feet) long
1 box*
paper about 8 cm × 8 cm (3'' × 3''), two per student
1 student magnet (see Appendix
string
iron straight pins

What to Do

1. Distribute the paper and have each youngster draw one object that a magnet can attract on one piece of paper and one that a magnet cannot attract on the other. Have them place a steel pin securely into the corner of the first paper. Place no pin in the other paper. Have them write their names on each drawing.
2. Make a magnet fishing pole (see illustration).
3. Have the children fish one at a time and show what they caught. Have them decide whether a magnet would be attracted to such an object. Some disagreement may occur on drawings of keys, TV cabinets, etc. Point out that a magnet is attracted to iron or steel and not to any particular structure. Therefore, iron keys will be attracted to a magnet but brass keys will not.

*Although this lesson is an extension of Lesson 3-3, it can be easily modified for other lessons such as fishing for verbs, pronouns, etc.

Building a Parachute

Specific Concepts/Skills

1. A parachute uses air pressure to help resist the force of gravity.
2. This lesson will help students learn to follow directions and teach them to build a model parachute.

MATERIALS

Per student
1 ''parachute paper''
4-8 glue-backed hole reinforcers
4 pieces string about 25 cm (10'') long
2-3 washers
1 figure made of paper, clay, etc. (see instructions)

VOCABULARY
model, parachute

Introducing the Activity

1. Review the previous class study of gravity and the forces that work against it.
2. Ask: "What things use air pressure to overcome gravity?" (An airplane, a glider, a balloon, birds, etc.)
3. Say: "We are going to build a model of something that allows a person to fall through the air and come to a soft landing on the ground. In the model, air pressure works against the pull of gravity. Who knows what we will build a model of?" (A parachute.)
4. Demonstrate how to build the model parachute and then give out the materials.

Instructions for Building a Model Parachute

a. Obtain a soft tissuelike paper. Apples and other fruits are shipped to supermarkets wrapped in such paper. Most produce managers will be happy to give it to you. A 30 centimeter (12 inch) square of wrapping tissue paper will do well.
b. Punch a hole in each corner of the paper with a pencil point.
c. Reinforce the hole with glue-backed hole reinforcers. For extra strength, use one on each side of the hole.
d. Cut four strings about 25 centimeters (10 inches) in length. Tie one string through each hole. Tie the loose ends of the strings together around the washer. Be sure the string lengths are equal.
e. Embellish the parachute by adding a paper, cloth, or clay person. A paper person can be attached to the washer by using a string and hole reinforcer. Punch a hole in the figure's head, stick on the reinforcer, and tie a short string through the hole and to the washer.
f. A small hole cut into the center of the parachute seems to help it open more quickly.

Glue hole reinforcers to front and back of paper.

12" sq. of tissue paper

This hole helps the parachute open quickly.

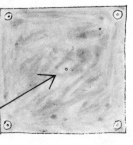

For this variation cut tissue square into this shape.

washer

paper

5. When the model parachutes are completed, they should be folded in preparation for launch. The simplest way to launch is to throw the folded parachute as high as possible. When it starts to fall, it should open and float gently down. However, if the parachute is folded poorly, it won't open and the poor model figure will be squashed. Allow the youngsters to determine the best way to fold the parachute. Launching is best done in the play yard.
6. Relate the parachute to gravity, air, and air pressure by asking questions such as:
   a. If a space man tried to launch your parachute on the moon, what would happen to it? (It would fall down unopened because there is no air to open or support the parachute.)
   b. Why did the unopened parachutes fall faster than the opened parachutes? (The air caught under the open parachute tended to keep it up and helped slow its fall.)
   c. Did the air pressure help the parachute overcome the force of gravity? (No, the parachute fell slowly, but it fell.)
   d. Do you think the size of the parachute would affect how fast the parachute will fall? (Yes.)
   e. Why did we need a washer? (Stabilizing force.) Will the parachute work without it? (Try.)
   f. Why do parachutists often carry two parachutes? (In case one doesn't work.)

Investigations at Home

Ask the class to design and build the very best model parachute they can. Give them what supplies you can spare, such as string and washers. Any light materials such as silk or cotton cloth can be used to build a more effective parachute. When they are returned to class, launch them from the highest available place, *e.g.*, a roof, second-story window, or private plane.

Evaluation

1. Were most children able to build their parachutes without help?
2. Were most children able to relate the parachute to air, air pressure, and gravity?
3. Did they want to take their parachutes with them when they left school? Did you let them?
4. Did one-fourth of the class return with home-designed parachutes?
5. This lesson was fun. Did everybody have a good time?

MATERIALS
Per team
1 clear plastic cup or MPC
3 different-sized rubber bands
extra rubber bands
tape recorder (optional)

VOCABULARY
pluck, vibrate, vibration, pitch

Investigating the Sounds of Student-built Musical Instruments

Specific Concepts/Skills

1. A vibrating object can produce sound.
2. The faster an object vibrates, the higher the sound, or pitch.
3. The slower an object vibrates, the lower the sound, or pitch.

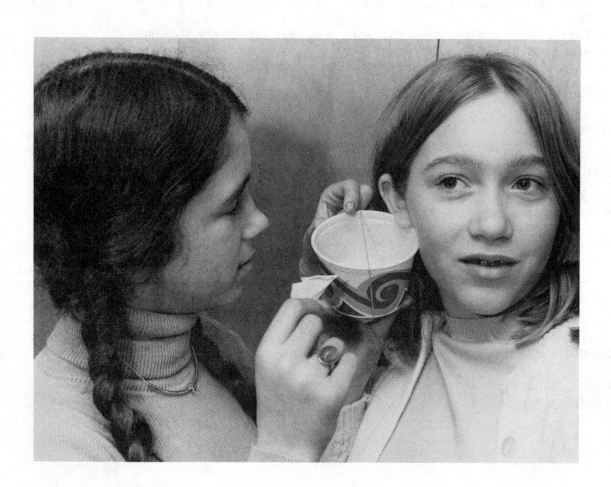

Note to Teacher

Lessons 3-5 through 3-8 are concerned with different aspects of sound. Consideration should be given to using them consecutively.

One successful way to focus the class's attention is to introduce each lesson with the Mystery Sound of the Day. This requires a tape recorder. Prerecord the mystery sound. For example, place the microphone in a can and drip water on the outside of the can. This will make an eerie sound. Or record the sounds of an auto exhaust pipe, a squeeky shoe, or a closing door.

This Mystery Sound will start the class thinking about and listening to sounds. Incidentally, you might award prizes, including such fabulous things as a genuine, wooden metric ruler (8 cents), an authentic 1966 D penny, an instant pumpkin seed (just plant and water and in about a year—instant pumpkin).

The Activity

1. Pass out the materials.
2. Direct each child to place any one rubber band around their cup. (Demonstrate.)
3. Say: "Place your ear near the cup. What do you hear?" (Nothing.)
4. Say: "Now, pluck the rubber band with your finger." (Demonstrate.) "Do you hear something?" (Yes.)
5. Ask: "In order to make sound, what must the rubber band do?" (Move back and forth.)
6. Say: "This quick back-and-forth motion is called *vibration*. When you pluck the rubber band can you see it vibrate?" (Yes.)
7. Say: "Let us see if you can transfer the vibration from the rubber band to a piece of paper. Two scientists will have to work together to do this."
8. Direct one young scientist to pluck the rubber band and the other to touch the tip of a piece of paper to the rubber band. (The paper will vibrate and give off a buzzing sound.)
9. Ask: "What did you hear?" (The paper buzzed.) "What caused the paper to vibrate and make sound?" (The vibrating rubber band.)
10. Review that an object that vibrates may make sound.
11. Direct the children to perform the next investigation. (They may have to work in pairs.) "Pluck your rubber band. Now, as you pluck it with one finger, move a finger of your other hand along the rubber band." (Demonstrate.) "What change occurs in the sound?" (It gets higher.) "Why does it get higher?" (When you prevent the whole rubber band from vibrating, the remaining part vibrates faster. The faster the vibration, the higher the sound, or *pitch*.)
12. Review: the faster the vibration, the higher the pitch, or sound; the slower the vibration, the lower the pitch, or sound.

13. Say: "Next you are going to build your own musical instrument."
14. Direct the class to carefully place the other rubber bands around the cup. (Demonstrate.)
15. Say: "Pluck the rubber bands and listen to the pitch. Decide which band has the highest pitch, next highest, and lowest."
16. Say: "Arrange the rubber bands from highest to lowest pitch. Try to pluck a tune."
17. Ask: "How do you think you could improve on your musical instrument?" (Let the discussion proceed freely, but reinforce the concepts of vibration and pitch, the highness or lowness of sound.)
18. Assign home investigation.

Investigations at Home

1. Have the youngsters build their own musical instruments and bring them to class. Have each youngster who brings one to class demonstrate how it works. Encourage them to use the words *vibration* and *pitch*. If possible, record the sounds of the better instruments by placing the microphone very close to them.
2. Ask those that play instruments to bring them to class and demonstrate.

Evaluation

1. Have most youngsters incorporated the words *vibrate* and *pitch* into their working vocabulary?
2. The investigations required manual dexterity. Were the youngsters able to perform the required activities? Did they cooperate?
3. Did they bring their homemade instruments to class?
4. Did they have fun? Did you?

MATERIALS
Per class
1 pair scissors
straws

Making a Straw Change its Pitch

Note to Teacher

This investigation is interesting, but it's hard to find someone who can do it.

What to Do

1. Ask: "Is there anyone in the class that can blow through a straw and produce a sound?" (If you can do it, demonstrate. The following may help: Flatten one end of the straw. Cut as indicated in the illustration. Place the cut end in your mouth, and blow into it while vibrating your pursed lips.)
2. If you have enough straws let everyone try—usually one or two can do it. If no one can get a sound from the straw, the investigation will have to await the discovery of a student or teacher who can.
3. Once the "straw musician" is found, have him produce the sound in front of the room.
4. Ask: "What would happen to the pitch if I cut off the end of the straw?" (The straw would be shorter and therefore vibrate faster. The pitch would get higher.)
5. After the discussion and student predictions (don't give the answer), snip about 2 centimeters (1 inch) off the straw. Be sure everyone hears the change in pitch. Then rapidly snip off pieces of the straw as the volunteer continues to blow. The pitch will get progressively and rapidly higher. A piercing cry means you have cut too much.
6. Repeat and review the relation between vibration and pitch.

MATERIALS
Per class
1 balloon

VOCABULARY
vocal cords

Investigating How Speech, Humming, and Whistling Sounds Are Made

Specific Concepts/Skills

1. Air moving past the vocal cords produces sounds.
2. The tongue, teeth, and lips help form the sound into words.

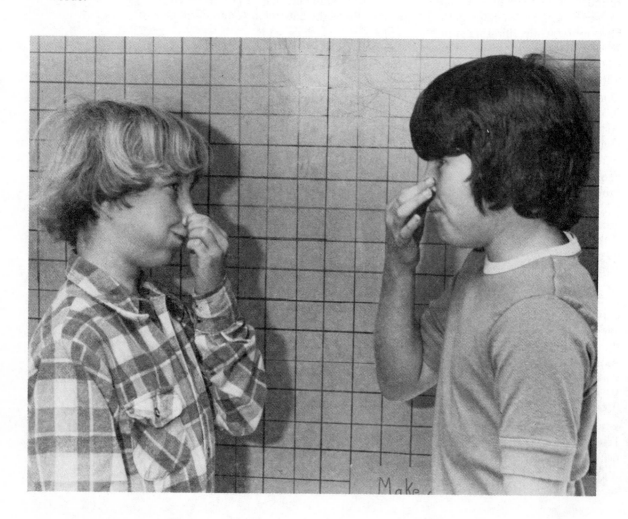

The Activity

1. Say: "We have learned that sounds can be made by vibrating objects." (Review Lesson 3-5.)
2. Ask: "You make sounds when you talk or hum. Do *you* vibrate?" (Allow discussion.)
3. Say: "Place your hand on your throat and hum. Do you feel vibrations? What do you think is vibrating?" (Air from the lungs passes between the vocal cords. They vibrate, causing the air to vibrate. We detect the vibrating air as sound.) You may liken the vocal cords to rubber bands being "plucked" by moving air.

    You may wish to demonstrate how moving air can cause something to vibrate and make sound. Blow up a balloon. Pinch the mouth end closed, then spread it, and let the air out slowly. Sound is produced, the pitch of which can be varied by changing the tautness of the opening. Liken the end of the balloon to the vocal cords, and the moving air to air leaving the lungs. The sound is made by the vibrating open end of the balloon (vocal cords).
5. Say: "Now, hold one hand on your throat and the other in front of your mouth and say a few words to your partner."
6. Ask: "Do you feel vibration?" (Yes.) "What is vibrating?" (Vocal cords.) "Did the hand in front of your mouth feel anything? (Yes.) "What did it feel?" (Puffs of air.)
7. Bring out in discussion that the air leaving the mouth first passed through the vocal cords and caused them to vibrate. And then, as the vibrating air passed the tongue, teeth, and lips, the sounds were formed into words.
8. Ask: "When you hum, does air leave your mouth?" (Allow them to try to figure it out. When you hum it is very difficult to feel the air leaving the nose. But, since air must move past the vocal cords to make them vibrate, and humming is done with a closed mouth, the air must be leaving via the nose.)
9. After discussion and experimentation on *the youngsters' part,* suggest that they close their mouths tight and hold their noses and hum. (The mouth quickly fills with air and they have to let it out or stop humming. However, if they swallow the air, they can continue humming.)
10. Ask: "When you whistle, do your vocal cords vibrate? Hold your neck and whistle. Do you feel vibration? (No.) "To have sound, something must vibrate. What do you think is vibrating here?" (Bring out in discussion that if they open their lips and try to whistle they can't. But the moving air causes their pursed lips to vibrate and make sound.)
11. To review what they have learned about sound so far, you may wish to ask the following question and encourage discussion:

    a. How does a mosquito make sound? (Vibrating wings.)

b. Whose wings move faster—a mosquito's or a bee's? (A mosquito's, whose pitch is higher.)
c. Does sound travel through air? (Yes, otherwise we couldn't hear each other talk.)
d. Whose vocal cords usually vibrate faster—a man's or a woman's? (A woman's. Since a woman's voice has a higher pitch, the vocal cords must vibrate faster.)
e. Some houses seem to make noise at night, but not during the day. Can you explain this? (Many young children are afraid of the sounds they hear at night. By bringing this question up in class, you can help dispel their fears. All sounds require some vibration. Many homes settle, and beams, etc., move and make sounds. This happens during the day, but is masked by other louder sounds.)

Investigations at Home

1. Tell the youngsters to practice imitating the sound of some animal or object at home. Then, in class the next day, have each youngster perform the imitation and see who can tell what it is supposed to be; e.g., dog, cat, wind howling, door closing, train.
2. Have youngsters bring anything to school that makes a sound and explain to the class what is vibrating and making the sound. (A whistle has a moving ball, the radio a moving speaker, etc.)

Evaluation

1. Were most children able to give *reasonable* answers to the review questions?
2. Did they cooperate during the investigations?
3. Did at least one third of the class do their home investigations?

MATERIALS
Per team
2 paper cups
450 cm (15 ft) of string per team
2 paper clips

VOCABULARY
(review) vibration, solid,
liquid, gas

Making a String Phone

Specific Concepts/Skills

1. Sounds can travel through solids, liquids, and gases.
2. Sounds travel best through solids, next best through liquids, and least well through gases.
3. A vibrating string can carry sound. When the vibration is stopped, the sound is stopped.

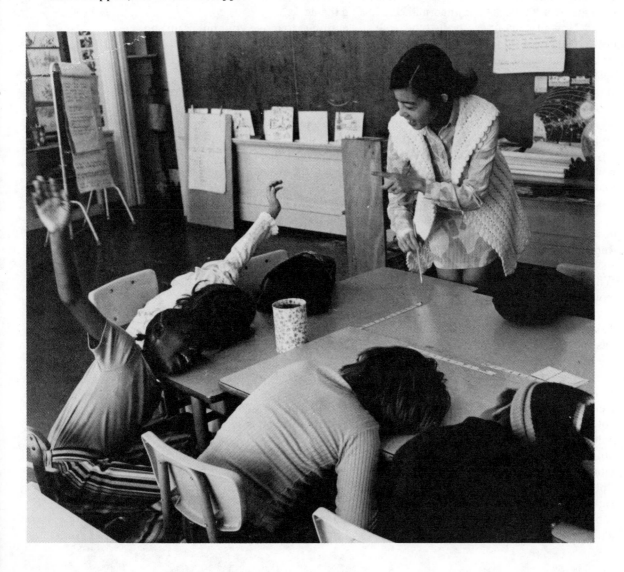

The Activity

1. Review: "We have learned that sounds are made by vibrating objects. We have also learned that sound can travel through the air. (You can hear me.)"
2. Ask: "Do you think sound can travel through your desk?"
3. Say: "Let's find out. Everyone must be very quiet during this investigation. Move your fingers up and down against your desk. Do you hear a sound?" (Some may, but most won't unless their fingernails are hitting the desk.)
4. Say: "Now place your ear against the desk and continue to move your finger up and down. Can you hear a sound." (All should.)
5. Ask: "Does sound seem to travel better through your solid desk or the air?" (The desk.)
6. If the desks are movable, try the next investigation. Place all the desks in one row together so they butt against each other.
7. Say: "Let's see if sound will travel from one desk to another." Have the first youngster in that row tap his pencil eraser against his desk.
8. Say: "Raise your hand if you can hear the eraser tapping against the desk." (Very few in the class should hear it. If they do, press the eraser against the desk rather than tapping it.)
9. Say: "Those of you in this row (with the desks together) place your ear against your desk. Raise your hand if you can now hear the sound of the eraser touching the desk." (All in that row should hear it, but the other class members won't.)
10. Repeat with each row of students.
11. Ask: "What would happen if the second desk in the row was not touching the first? Why?" (The path of the sound would be broken and those at the remaining desks would not hear the sound.)
12. Try it.
13. Ask: "Do you think a string will conduct sound?"
14. Pass out the strings. Stretch the strings out. Have one partner pluck the string while the other listens. Then have them reverse roles.
15. Say: "We can improve on the way the string delivers sound by adding cups to each end."
16. Demonstrate how to build a string phone. Thread the string through a hole in the bottom of each paper cup. Tie the end of the string around a paper clip so it won't pull out of the hole.
17. Stretch the string tight. (Prevent it from touching anything, as the string would not vibrate beyond an obstruction.) Have one youngster talk very quietly into the "phone" while another listens. They should be able to hear only each other.
18. Have each team build a phone and test it in class or at recess.
19. A phone tap can be made by crossing one set of phone lines over another (see illustration). You may want to discuss the morality of tapping a neighbor's phone.

Tap is made by looping string around another phone line.

20. If you have a headache, you might want to send the class outside at the end of a 150-meter (500-foot) string phone.

Investigations at Home

1. Many youngsters may want to continue to experiment with the phones in class. Suggest rather that they build the very best string phone they can at home, trying various kinds of strings and phone ends. They can demonstrate their string phones in class. The best one will conduct sound over the longest distance most clearly. The tests may need to be made in the hall or outside.
2. Ask the youngsters to perform an investigation that will prove whether sound is conducted in water (it is). They might place an ear into a sink filled with water and then tap two spoons together in the water. The sound will be more clearly heard in water than in the air above it.

Evaluation

1. Do all the youngsters know that sound travels through solids, liquids, and gases (air being the only gas tested)?
2. If the sound did not travel from one phone to another, were the team members able to discover why, or did they wait for your solution?
3. Did you solve their problems or encourage them to solve them? (The usual problem is that the string touches something so that the vibration is absorbed.)
4. Did at least one third of the class do their home investigation?
5. Could any of the youngsters explain why the cup was needed?
6. Have the youngsters improved their ability to work with the science materials and with their partners?

MATERIALS
1 electric vacuum pump and
    stand
1 bell jar with electric bell
1 jar vacuum grease
rubber tubing
6V cell and connecting wire

VOCABULARY
vacuum

The Sounds of Silence

Note to Teacher

The electric vacuum pump can be one of the most useful pieces of science equipment
for teaching beginning science. Unfortunately, most elementary schools do not own
one and, therefore, this investigation is optional. However, junior and senior high
schools usually do have the pump and related accessories, and by contacting an
understanding teacher, you may be able to borrow what you need. After all, the better
you prepare your students, the better their students will be prepared.

What to Do

During this investigation an electric bell is going to ring in a vacuum jar. As it continues
to ring, the air will be pumped out of the jar. Although the bell continues to function, no
sound can be heard, proving that sound cannot travel in a vacuum.

Set up the equipment as illustrated and explain what it is and how it functions. Have a
student turn on the bell by connecting the wires to the battery terminals. Be sure

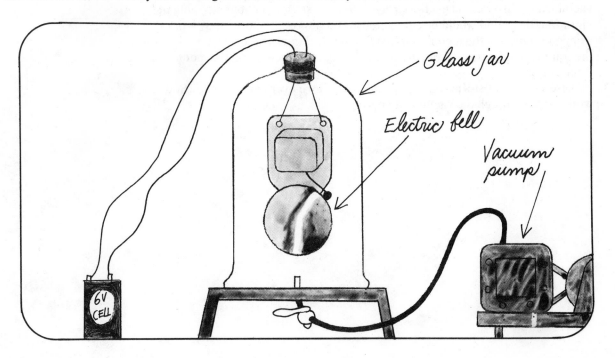

Glass jar

Electric bell

Vacuum pump

6V CELL

everyone hears and sees the ringing bell. Shut it off. Then ask what would happen if you turned on the vacuum pump and removed the air from the vacuum jar. That is, can sound travel in a vacuum?

Allow time for discussion. Review that vibration is necessary for sound and that sound can travel through a solid, liquid, or gas. A vacuum is none of these things. Have all class members commit themselves to a prediction as to what will happen when you turn on both the bell and the vacuum pump.

Turn on the bell first, then the pump. As the air leaves the vacuum jar, the sound will gradually get softer until it cannot be heard. Point out (if necessary) that the hammer on the bell is still moving. Shut off the pump. The vacuum will remain. Then turn the valve and allow air to gradually enter the vacuum jar. The sound will gradually return.

Repeat the original question: Can sound travel through a vacuum?

You may wish to extend the discussion with questions such as:

1. "If two satellites crashed into each other in space, would the sound be heard?" (No.)

2. "If two space stations on the moon could not use their electric communicators, could they set up string phones and use them inside one station?" (Yes, if the strings didn't touch anything and there was air in the station.) "Could they use string phones from one station to the next?" (Possibly; the string will vibrate in a vacuum and if they could get the strings into each station without stopping the vibration, the sound would be conducted in the station.) "Would they work if used from one station to the moon's surface?" (No. Although the string would vibrate, the sound couldn't travel from the string through the vacuum on the moon to a listener's ear.

3. "There are violent storms on the sun. Why can't we ever hear them?" (Sound won't travel through the vacuum of space.)

MATERIALS
Per class
1 roll of thread
tape recorder (optional)
Per student
1 metal hanger
2 telephone cups

VOCABULARY
stereophonic sound (optional)

Making a Stereo Metal Hanger Musical Instrument

Specific Concepts/Skills

1. Sound will travel from a vibrating metal hanger, through a string and the fingers, into the ears.
2. Stereophonic sound is sound heard from two locations.

Note to Teacher

Prepare for this lesson a day in advance by asking *each* youngster to bring a metal coat hanger to class. Tell them that they will use them to build a musical instrument that only they can hear.

The Activity

1. Demonstrate how to build the instrument. Tie a thread or string about 45 centimeters (1-½ feet) long around each end of the hanger. That's it; you have built it! Wind a few inches of the end of one thread around the top of one index finger, wrap the other thread around your other index finger and place the fingers in your ears. Tap the hanger against the desk and you will hear a symphony with overtones and vibrations that fill your head with stereophonic sounds that no one else can hear.
   Note: The sound is *greatly* improved by knotting the ends of the strings into the cups used to make a string phone.
2. Ask the youngsters to build their own. Supply thread as needed.
3. Ask the youngsters to experiment with their instruments as follows:

   a. Have someone tap the hanger with a pencil or other object in different places and note if the sound changes.
   b. Remove one finger from an ear and note the change in sound (you lose the stereo).
   c. Have someone hold the thread after you strike the hanger. What happens to the sound? Why? (You can't hear it because the string's vibration is stopped.)
   d. Change the shape of the hanger; open it and bend it as you will, and note the changes in sound.
   e. Replace the hanger with other objects such as hair clips, metal combs, any thin metal grillwork, paper clips, and note the sounds.
5. Ask: "How do you think this musical instrument works?" (When you strike the hanger, it vibrates in many different ways. The various vibrations are transmitted to the strings and conducted along them to your ears. The vibrating string causes the air in your ear to vibrate and produce the sounds you hear with your brain.)

Investigations at Home

If there is still lots of interest, ask the children to make the best-sounding instrument they can and bring it to school. Tell them they can use anything and get anyone's help. If a tape recorder is available, try to record the sounds so the entire class can hear (it's difficult, but placing the microphone in the cup helps).

Evaluation

1. This lesson was fun, did the class have fun? Did you?
2. When asked to explain how the instrument worked, were many of them able to utilize their previous experience with sound and sense investigations?
   Note: The stereo sound can be remarkably improved by using the telephone cups (see Lesson 3.7) at the end of the string.

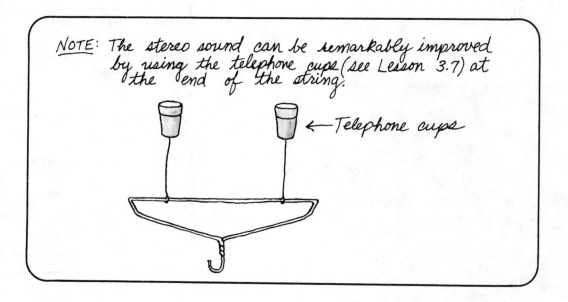

NOTE: The stereo sound can be remarkably improved by using the telephone cups (see Lesson 3.7) at the end of the string.

← Telephone cups

Batteries, Bulbs, and Circuits

Specific Concepts/Skills

1. A battery is a source of electricity.
2. A circuit is the path electricity travels from one terminal of a battery to the other. A bulb may be part of a circuit.
3. A short circuit is made by connecting one terminal directly to the other. A short circuit will soon wear out a battery.

MATERIALS

Per class
large drawing of a bulb and battery (prepared in advance or drawn in class on the board)
Duplicated sheets of student work
knife or can opener

Per team
1 battery (D cell)
1 2-cm × 10-cm (5″ × 1″) strip aluminum foil
1 #47 or #48 bulb (These 6-volt bulbs last longer, but won't get very bright with one D cell. If you can afford it, buy the 3-volt bulbs used in flashlights instead.)
1 rubber band (optional)
2 pieces of masking tape
1 magnifying lens

VOCABULARY

battery, bulb, positive (+), negative (−), terminal (optional), circuit, short circuit, filament, incomplete circuit

Filament
Glass bulb
Ceramic
Wire
Solder spot
Base
Insulation
Metal lip is conductor

Note to Teacher

Lessons 3-9 and 3-10 are geared toward youngsters with little experience with electricity and its uses. The recommended materials are very inexpensive; the concepts developed, quite basic, and the investigations, easy and fun to do. However, young children may have difficulty understanding the concepts. Your guidance and patience will be required. Magnetism is studied in Lessons 3-12 and 3-13. The relationship between magnetism and electricity is investigated in Lesson 3-14.

Caution: Emphasize clearly that the electric outlets found at home can be dangerous and that investigation should be confined to batteries.

The Activity

1. Pass out the materials to each team.
2. Have the youngsters describe the battery and make a drawing of it. You may wish to guide them with questions such as:

   a. Does the battery have a top and a bottom? How can you tell? (The writing and flat bottom indicate top and bottom.)

   b. Which side is positive (+) and which is negative (−)? (It's marked. The terms *plus* and *minus* may be used with younger children.)

   c. What is the top and bottom made of? (Iron or steel. A magnet is attracted.)

   d. What are the sides made of? (Cardboard. Use a knife or pointed can opener to scrape back the covering, exposing it as cardboard.)

   e. What does it say on the battery? (Size [D], volts [1.5], name, etc.)

   f. For what is a battery used? (Source of electricity.)

   g. Where have you seen batteries used?

3. Have the youngsters describe the bulb and draw its external and internal parts. The hand lens is useful for observing its small parts. You may guide them with questions, such as:

   a. What is the bulb made of?

   b. Is the small wire (filament) connected to the two larger wires? (Yes.)

   c. What is the shape of the small wire? (That of a spring.)

   d. Do the two large wires touch each other? (No; connected only by the small springlike wire.)

   e. Where are the large wires attached in the base of the bulb? (This is difficult to see, even with a hand lens. To aid the class, you can break one small bulb by placing it under a piece of paper and hitting it with the battery. Or demonstrate with a clear 60-100W bulb. If a clear bulb is unavailable, place a frosted bulb in a paper bag and break it with a battery or hammer.) One wire will be found connected to the metal base. The connection, marked with a solder spot on the base, can be easily seen. The other wire is connected to the tip of the bulb. The tip is separated from the metal base by insulating material. Thus, the two large wires are connected only by the filament.)

4. Say: "Using only the aluminum foil, battery, and bulb, make your bulb light (see Note to Teacher). "Each time you try a method, even if the bulb does not light, make a drawing of the connections you make." (When a youngster finds one method that works, ask him to find another. Sharp youngsters may be challenged by providing them with a piece of foil too short to touch both ends of the battery. Allow them to discover the need to tear and splice the foil to obtain a sufficient length. To assure tight connections, one member of the team can squeeze the foil to the battery or masking tape or a rubber band may be used.)

5. After allowing sufficient time for trial-and-error investigation, have a few successful youngsters place their drawings on the board. Have several "unsuccessful" youngsters place their drawings on the board too. Point out that knowing an unsuccessful method is useful since it helps you redirect your time and effort.

6. Ask: "Why was one method successful and not another?" Using a large drawing, show the need for the electricity to travel in a complete path, or *circuit*. If the circuit is not complete electricity will not flow. The circuit must lead from one end of the battery (terminal) to the bulb through the filament (which gets hot and gives off light) and back to the other end of the battery.

   Many unsuccessful youngsters will have made a complete circuit but will have bypassed the bulb. This is a *short circuit*. Point out that in a short circuit the electricity flows very quickly from one terminal to the other and soon the battery will no longer supply electricity. The aluminum foil may become warm in a short circuit. This is caused by the relatively rapid and abundant flow of electricity.

7. Using worksheets or board illustrations, have the youngsters predict what circuits will light the bulb and why. Allow them to test their predictions by setting up the circuit.

Evaluation

1. Did the youngsters show increasing ability to observe and report accurately?
2. Did the youngsters understand the value of reporting successes and failures when trying to get the bulb to light?
3. Do at least half the youngsters understand the meaning of *complete circuit* and *short circuit?*
4. Can at least half the youngsters trace, on an enlarged drawing, the path of electricity from the battery through the bulb and back to the other end of the battery?

STUDENT WORKSHEET
Predict which circuits will light the bulb, then try it.

Name _____

1 It will light _____
It won't light _____
Did you predict correctly _____
Why? Incomplete circuit _____
Short circuit _____
Complete circuit _____

2

3

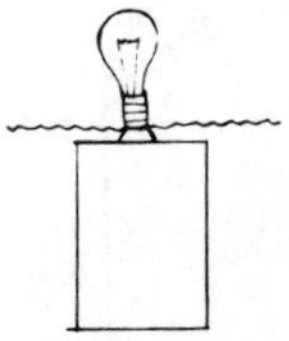

4

5

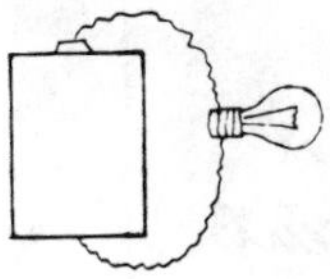

6

7

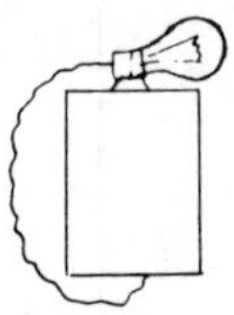

8

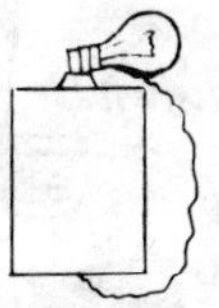

9

Draw your own circuit

Draw your own circuit

Draw your own circuit

Conductors and Insulators

Specific Concepts/Skills

1. Electricity will travel through materials known as conductors. Metals are good conductors. Nonmetals are poor conductors.
2. Some poor conductors are used to insulate metal wires.

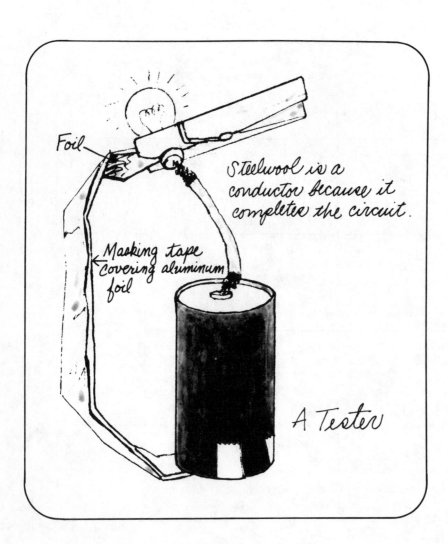

Foil

Steelwool is a conductor because it completes the circuit.

Masking tape covering aluminum foil

A Tester

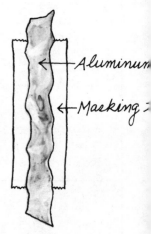

Aluminum

Masking

Insulated Wire

MATERIALS
Per student or team
Needed to build tester:
1 D cell
1 strip aluminum foil about 2 cm × 15 cm (1″ × 6″)
1 #47 or #48 bulb
1 clothespin
masking tape (about 15 cm [6″ long)

Things to test:
15 cm (6″)	paper clip
insulated wire	coin
steel wool	wax paper
rubber band	pencil
clear tape	etc.
cord	

To make insulated wire:
masking or clear tape
aluminum foil
plastic bag for prepacking (optional)

VOCABULARY
conductor, insulator

Note to Teacher

The three student tasks included in this lesson are:

   a. Building a device to determine if something is a conductor. This device is simple to build, helps the youngsters understand the concept, and costs almost nothing. Bulb holders that are easier to use are available but cost up to 50 cents each and obscure the concept somewhat.

   b. Testing to determine conductors and nonconductors.

   c. Making two kinds of insulated wire.

The Activity

1. Review Lesson 3-9.
2. Say: "We have seen that electricity will move from one terminal (end) of the battery to the other along aluminum foil. The aluminum foil is known as a *conductor.* It conducts electricity."
3. Ask: "What other things will conduct electricity?" (Allow for discussion and suggestions.)
4. Say: "Let us build a device that can be used to find out what things conduct electricity."
5. Pass out the materials. (It saves time to have them pre-packaged in plastic bags.)
6. Demonstrate how to build a tester (see illustration):
   a. Fold a 15 centimeter (6-inch) strip of aluminum foil lengthwise until it is about a centimeter (a half inch) wide. Cover the foil with a strip of masking tape. Allow the ends of the foil to protrude.
   b. Place the base of the bulb into the clothespin.
   c. Snuggle one end of the aluminum foil against the base of the bulb. The clothespin should hold it in place.
   d. Tape the other end of the aluminum foil securely to the positive (bumped) end of the battery.
   e. Test the circuit by touching the tip of the bulb against the negative (depressed) end of the battery. The bulb should light.
7. Ask: "How can we use the tester to find out what things conduct electricity?" (Allow the youngsters to experiment. They should discover that they can touch one end of the item to be tested against the negative terminal of the battery, and the other end against the tip of the bulb. If the item is a conductor, the bulb will light (see illustration).

Testing to Determine Conductors and Nonconductors

8. Allow youngsters to test each item and others they may select. Have them place conductors in one pile and nonconductors in another.

9. Ask: "Do the things that conduct electricity have anything in common?" (All are metal.) "Do the nonconductors have anything in common?" (Nonmetals.)Some nonconductors are used as insulators.

10. Say: "A metal wire is usually covered with an insulator." (Have the youngsters examine the wire they tested.)

Making Insulated Wire

Say: "We have found that the steel wool is a conductor and the masking tape a nonconductor. Find a way to combine the two and make an insulated wire. Test the wire to be sure it conducts electricity while the insulator does not (see illustration). Caution: Use no more than one battery. With more than one, the steel wool sparks and burns.

Say: "Use masking tape (or clear tape) to insulate some aluminum foil. Test it to be sure the wire conducts and the tape insulates."

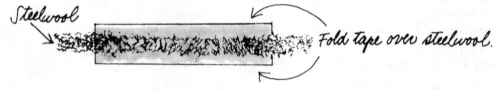

11. Review and apply what they have learned by asking questions such as:

   a. "Why are wires insulated?" (To prevent short circuits that may be caused if two wires touch; to allow people to touch wires conducting electricity without getting a shock; to prevent a wire that gets hot [due to electricity passing through it] from burning its surroundings.)

   b. "What kinds of insulators are used on house wires?" (Rubber, plastic, waxed cord. If possible have samples to demonstrate.)

   c. "What metal is used most often to make wire?" (Copper.)

   d. "One of the best conductors is silver. Why isn't silver often used as wire?" (Too expensive.)

   e. "Is the filament of the bulb insulated?" (No.) "Why not?" (It is supposed to get hot and give off light.)

   f. "Does the bulb contain insulation?" (Yes, the disk between the wires is insulation, and the wire that extends to the top is insulated from the metal base.)

Investigations at Home

The making of insulated wire can be a home investigation. The testing can be done in class or at home. Suggest they make wires of various flexible conducting materials such as solder, paper clips, or hair clips.

Evaluation

1. Even though you cautioned against it, did a youngster try to light his bulb by placing his wires into a wall socket? (That has happened in several classrooms!) Does he listen more carefully now?
2. Was the classroom busy and generally quiet?
3. Could all or most youngsters build a tester?
4. Have you been asked lots of questions you couldn't answer? Did you:

   a. Answer them anyway!
   b. Say: "That's a good question, why don't you look up the answer and report to the class?"
   c. Ignore questions you couldn't answer.
   d. Admit you didn't know the answer and let it go at that.
   e. Admit you didn't know the answer, decide if the answer could be found and understood and then find ways to discover the answer, through investigation, library research, or appeal to an authority.
   f. All of these at different times.

MATERIALS
advertisements for batteries
fresh samples of batteries of
  various brands, all of the same
  size and voltage
simple bulb circuit for each
  battery

VOCABULARY
variable

Finding Out Which Battery Lasts Longest

Note to Teacher

This investigation is to determine which battery lasts longest. It can be introduced by showing the various advertisements, claiming long life for battery X, Y, or Z. Discuss how the class could find out which battery really does last longest. Cost is high, and therefore actual investigation should perhaps be left to one team.

What to Do

One method of investigation is to place different batteries into a simple bulb circuit. Keep all things (*variables*) the same, except for brand of battery. Place each battery in an MPC and close all circuits at the same time. Before class is dismissed, open the circuits. Resume the next class day. Possibly add such extras as:

a. How many hours does each battery last?
b. Which battery gives the most for the money?
c. Would the same results be found a second time?
d. Send the results of your tests to each manufacturer and ask for their comments.
e. Were all lamps as bright? Does that matter?

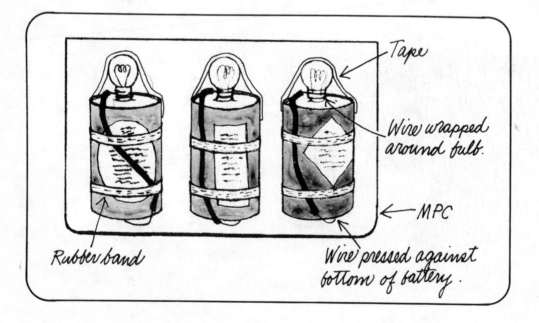

Building a Flashlight

Specific Concepts/Skills

1. There are many ways to solve a problem.
2. The application of prior knowledge, scientific methods, cooperation, and luck often proves effective in solving technological problems.

MATERIALS

Per class
Styrofoam cups
various kinds of wire
cardboard rolls from paper products
thumb tacks

Per student
1 D cell
1 #47 or #48 or 3-volt bulb
aluminum foil
masking tape

VOCABULARY

on-off switch, flashlight, reflector

Note to Teacher

The youngsters are told to build the best flashlight they can, and off they go. Your job is to supply material they request and that's all. All youngsters assume responsibility, apply what they've learned, try, fail, and try again; build the flashlight, and develop a good feeling for themselves—and you.

The Activity

1. Say: "Today I would like you to build the best flashlight you can. Be sure to include a reflector and an on-off switch. Each of you will be given a battery, bulb, some foil, and masking tape. If you need anything else come up and ask for it. If we have it you can borrow it. You may work alone or with your team members. If you decide to work together you can combine your batteries."
2. Pass out materials.
3. Relax. Allow youngsters to proceed without help.
4. As the flashlights are completed, you may wish to appraise them and have youngsters explain how they work. Completed flashlights may be compared for:

   a. style;
   b. ease of operation;
   c. effectiveness (how far and how wide the light is projected);
   d. durability;
   e. how well the on-off switch works.

Investigations at Home

Youngsters probably have some ideas that they couldn't apply in class. Encourage them to build an even better flashlight at home, and bring it to class for demonstration. (A problem may develop concerning materials and parental help. Some youngsters may not have access to batteries and you may wish to lend some out. Others will come in with a flashlight in part constructed by an adult. That's really no problem. In fact, the assignment will have stimulated communication between the youngster and his family.)

Evaluation

1. Did youngsters build different kinds of flashlights?
2. Did some cooperate, combining batteries to make more effective flashlights?
3. In discussing their work, were words such as *open* and *closed circuit* and *reflector* used correctly?

MATERIALS
Per team
1 magnet
2 pieces masking tape
1 toothpick
1 iron pin or thin nail
Per class
1 compass

VOCABULARY
attract, repel, north and south
pole, standard

Investigating the Poles of a Magnet

Specific Concepts/Skills

1. Magnets have two unlike poles. One is called north and the other south.
2. Unlike poles attract each other; like poles repel each other.

The Activity

1. Distribute the materials (save the pin or nail for homework).
2. Allow time for the youngsters to "play" with the magnets.
3. Ask: "Does your magnet attract or repel your partner's magnet?" (Discuss the meaning of *attract* and *repel*, and encourage students to use these words.)
4. Say: "One team member should lay a magnet flat on the desk. Place a piece of tape on the magnet (demonstrate). Place a toothpick through the tape and into the hole in the magnet.
5. Say: "Now the other team member should lower a magnet over the toothpick (see illustration on previous page).
6. Ask: "Do the magnets attract or repel?" (Either may occur.)
7. Say: "Repeat, turning the untaped magnet over. Now do they attract or repel? (They will reverse what they did before.)
8. Ask: "Are the faces of your magnet the same?" (No.)
9. Write the following on the board.

   Unlike faces attract.
   Like faces repel.

10. Discuss the meaning of these statements.
11. Say: "Place a piece of tape on the face of the untaped magnet that is *like* the taped face." (Since like faces repel, the face that repels the taped face is a like face.)
12. Have the youngsters show why they taped the magnet as they did. This tests their understanding.
13. Say: "The face of the magnet is also known as the pole. All magnet have two poles. The poles are usually found at the ends of a magnet, but the poles of our magnets are on the faces. (You may wish to draw or demonstrate bar and horseshoe magnets.)
14. Have more mature youngsters trade one magnet with another team. Ask: "Do the taped faces attract or repel?" Discuss the results. (They will vary since different magnets are being tested.)
15. Ask: "How can we arrange the magnets so that all taped faces are alike?" (Use one taped face as a standard for comparison with the others. Or a compass can be used. Whatever face the compass points to can be taped. All taped faces will then be alike (and, incidentally, the north pole face of the magnet.)

Investigation at Home

Distribute one iron pin or thin iron nail and a bit of Styrofoam cup to each student. Have them rub the pin across the face of the magnet 25 times in one direction. This will magnetize the pin. Have them test it to be sure it is magnetized. Tell them that at home they are to place the pin on the Styrofoam and the Styrofoam in a cup of water. Ask them to observe what happens and to attempt to explain why. (They have made a water compass. The pin will turn until one end faces north and the other south. However, if the pin is near iron, it will be attracted to that and will not point true north-south. If the youngster can get the pin to float without the Styrofoam, it will more readily respond to the earth's magnetic field.)

Magnetized pin on Styrofoam

Water in cup.

Evaluation

1. Were the youngsters able to find the like and unlike poles?
2. Were they amazed when one magnet floated in air? Were you?
3. Was the word *gravity* used by the youngsters? (One magnet overcame the force of gravity.)
4. Did at least half the youngsters do their home investigation?
   Note: The magnetized pin can be made to float on water without using Styrofoam. To do this just lay a small tissue on the water's surface. Place the magnetized pin on the tissue. Use a toothpick to gently submerge the ends of the tissue until it sinks, leaving the pin afloat. Why does the pin float on water? Because of the surface tension of the water.

   You can make the pin whirl by turning the magnet's face.

MATERIALS

Per team
20 paper clips or pins
2 paper clips
2 different magnets
graph paper (optional)
iron pin (optional)

VOCABULARY

measure, reliability, unit of measure, standard

Measuring the Strength of a Magnet

Specific Concepts/Skills

1. A reliable method of investigation can be repeated by others with consistent results.
2. The strength of a magnet may be measured using a standard unit of measure.

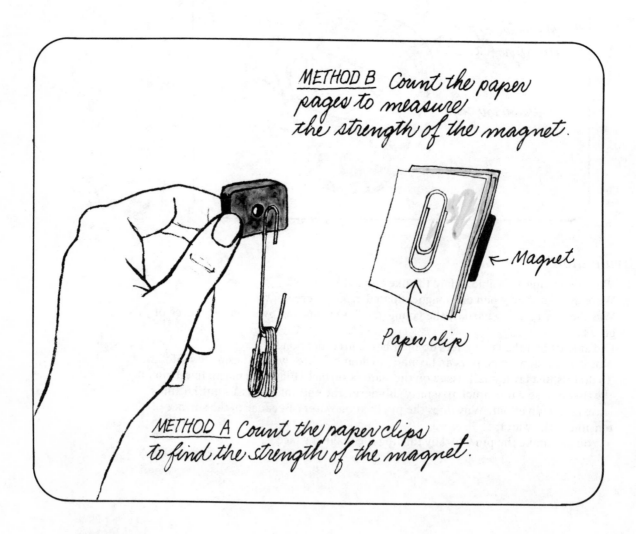

METHOD B *Count the paper pages to measure the strength of the magnet.*

← *Magnet*

Paper clip

METHOD A *Count the paper clips to find the strength of the magnet.*

Note to Teacher

During this investigation, youngsters seek a reliable method of measuring the strength of two different magnets. The concept of finding a standard (consistent) unit of strength should be further developed. The suggested units include the "paper clip," the "cm" and the "page."

The scientific unit normally used is called gauss. This unit is best left for later in the student's studies.

The Activity

1. Distribute materials.
2. Allow time for the youngsters to "play around."
3. Say: "You have had some time to examine the materials; now I would like each of you to answer this question: "*Which magnet is stronger*?)
4. Ask: "How do you know which is stronger?" (Allow discussion.)
5. Ask: "Can you find a way to *measure* how much stronger one magnet is than the other?" (Finding a solution to this problem is challenging. At first, youngsters may stick paper clips onto the magnet and compare the number stuck to each. This method uses the paper clip as a unit to measure strength, but the results are inconsistent and therefore unreliable. Discuss this and encourage youngsters to find more reliable methods. Allow a youngster with a promising method to show it to the class and encourage positive evaluation. Point out that often one idea may lead another person to an even better idea. Many scientific problems are solved this way.)

Suggested Methods to Measure Magnet Strength

A. One way to determine relative strength is to measure lifting capacity of each magnet. A simple device can be made using a paper clip as a hanger and other paper clips as units of weight.

First, locate the strongest surface of each magnet. Place the paper clip against that surface and hang paper clips on one at a time. If four stay on, but the fifth causes the clip to fall, the strength may be recorded as 4+ paperclips. The entire class can use this method with fairly consistent results.

B. A second method is to place the magnet on one side of a piece of paper and a paper clip on the other. Continue to add paper between the magnet and paper clip. When the magnet will no longer attract the paper clip, count the pages. Repeat with another magnet and compare strength in "pages".

C. A third method might measure pulling power along a horizontal surface in centimeters. Place the magnet on a piece of graph paper (or ruled paper). Determine maximum distance at which pin can be moved by magnet.

—Ruled paper

6. Review and amplify the concepts by asking such questions as:

a. How many paper clips does magnet A hold?
b. How many paper clips does magnet B hold?
c. How much stronger is A than B?
d. Did everyone get the same answers? Why not? (Magnets may not be the same strength, paper clips may not weigh the same, points of attachment of paper clip will vary, etc.)
e. Why must the paper clips be the same size and weight?
f. Could you have used some unit of measure besides the paper clip? Give examples. Would the results have been the same?
g. What would happen to the strength of the magnet if we used heavier paper clips? (Strength remains unchanged but the measure of strength will change, i.e., magnet A is 10 small paper clips strong or 2 large paper clips strong or 1 large and 5 small paper clips strong. The strength is unchanged; only the unit of measure changes.)

Investigations at Home

Encourage youngsters to think about other ways to measure a magnet's strength.

Evaluation

1. How resourceful were the youngsters in finding ways to measure the magnet's strength?
2. Do a majority of youngsters appear to understand that the unit of measure can be changed without affecting the strength of the magnet?
3. Was any youngster concerned that the two magnets were not the same size or weight and that this might be considered when measuring strength?

MATERIALS

Per student
20 **team** paper clips
paper clips (bent)
1 meter (3 feet) of narrow-guage
  winding wire
1 D cell
1 iron screw or nail, about 3-4
  cm (1½-2 in.) long
1 magnet
sand paper or emory board
wire stripper
1 compass (optional)

VOCABULARY

permanent magnet,
electromagnet, temporary
magnet

Building an Electromagnet and Measuring its Strength

Specific Concepts/Skills

1. Electricity and magnetism are related. When current flows through a wire a magnetic field is produced.
2. An electromagnet's strength can be varied by varying the length of winding wire or the strength of the current.
3. A temporary magnet can be made by stroking a permanent magnet.

Note to Teacher

During this investigation, youngsters make a temporary magnet and an electromagnet. The strength of the electromagnet is measured and then varied by changing the voltage (number of batteries) and length of winding wire. The strength of the electromagnet is measured by counting the number of paper clips or washers it can hold.

The Activity

1. Pass out the materials. (Pre-packaging will save class time.)
2. Say: "The magnets we have been using are known as *permanent magnets*." What does *permanent* mean? (Permanent magnets will retain their magnetic property for a long time. Heat, banging, or placing near stronger magnets, however, may cause them to lose their magnetic property quickly.)
3. Ask: "Do you think you can use your permanent magnet to change a washer (or paper clip) into a temporary magnet?" (This is a review problem of the Lesson 3-12 home investigation.) Allow time for investigation and give help if required. The washer should be stroked across the stronger magnet in one direction 15-20 times. The washer should then show weak magnetic properties. Discuss the meaning of *temporary magnet*.)
4. Ask: "How strong is the washer magnet?" (Encourage youngsters to measure its strength using techniques developed in Lesson 3-13. They will quickly discover that the magnet's strength is less than one paper clip. You might consider having some youngsters develop a more sensitive unit of measure that can be used, *e.g.*, pins.)
5. Say: "Using the materials you have, you can build an *electromagnet*."
6. Demonstrate how to build an electromagnet. Leave about 7 centimeters (3 inches) of wire free before and after winding. Wind carefully using the grooves in the screw as a guide. The wire is insulated. The insulation must be removed from about a centimeter (a half inch) of each free end. Use the sandpaper or emory board to do this with a lacquer-covered wire. A wire stripper is best with all other insulated wire. Some teachers prefer the youngsters to "discover" the insulation in trying to figure out why their electromagnet doesn't work.
7. Say: "Build an electromagnet. Predict how strong it will be (number of washers or paper clips it can hold) and then test it. (Caution youngsters not to keep the electromagnet on too long or the battery will soon wear out. A piece of masking tape can be wrapped around the wire to keep it from uncoiling.)
8. Have youngsters compare the strength of their electromagnets and explain differences. (You may wish to have them compare their electromagnets with the permanent magnets.)

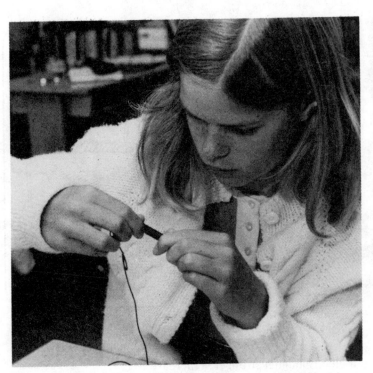
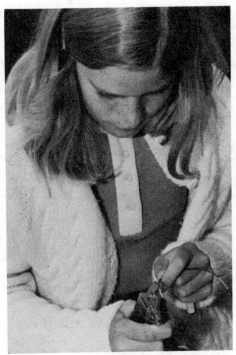

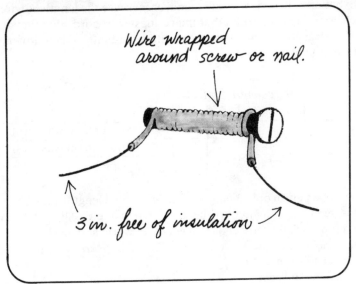

Wire wrapped around screw or nail.

3 in. free of insulation

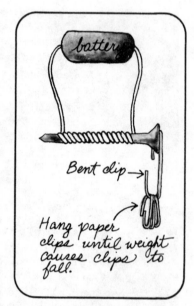

battery

Bent clip →

Hang paper clips until weight causes clips to fall.

9. Ask: "How can you make your electromagnet weaker? Try it." (Uncoil about 60 centimeters [2 feet] of wire and strength will be greatly reduced.)

10. Ask: "How do you think you can make your electromagnet stronger? (Increase voltage by using two batteries in series [front to end] or increase the length of wire wrapped around the screw. It is important that the youngsters record what they do to increase or decrease strength and measure the resulting strength. The following chart may be helpful in organizing data.)

	Wire length (in feet)					
	1	2	4	6	8	10
One battery						Strength (in washers)
Two batteries						Strength (in washers)

Since youngsters have been given only 3 feet of wire, additional lengths may be given to each. That is, team 1 gets a 4-foot piece; team 2, 6 feet, etc. All teams can double up batteries. However, the wire ends will get quite warm and masking tape should be used to hold the wire to battery terminals. The youngsters can squeeze the masking tape to make the connection tight.

11. Review by asking questions such as:

a. Would you use a permanent magnet or electromagnet to do the following and why?

   (1) Hold a note to your refrigerator. (Permanent; strength and variability of electromagnet not needed.)
   (2) Pick up spilled pins. (Permanent or electromagnetic, for quick release of pins.)
   (3) Move wrecked autos. (Electromagnet, for strength and ability to release.)
b. How may magnets be used to prevent cars from having rear-end collisions? (Tape the north face of a magnet to the front and rear bumpers of two toy cars. As they approach a rear end collision, the like faces will slow down the cars.)
c. Do electromagnets have poles? (Yes.)
d. What kind of magnet can be turned on and off? (Electromagnet.)
e. What is loadstone? How can one loadstone help make many magnets? (Naturally magnetic rock; by stroking.)
f. Can you build an electromagnet without the screw or nail? Try it. (Yes, but very weak.)

Investigations at Home

1. Provide volunteers with two batteries, a screw, and as much winding wire as you can spare. Have them build (and test) the strongest electromagnet they can. Have them demonstrate in class. (This is a relatively easy and yet fascinating assignment. Why not allow a youngster who needs an ego boost to do this task?)
2. Have volunteers use the principle of the electromagnet to build devices that employ one or more electromagnets such as a simple telegraph. The school library is usually a good source for projects.

Evaluation

1. Were all youngsters able to build an electromagnet?
2. Did the relationship between strength and length of wire and number of batteries become apparent?
3. Were they able to measure magnetic strength using washers as the unit of measure?

Conclusion

One of the great breakthroughs in science was the accidental discovery of the relationship between electricity and magnetism. Many books describe Oersted's serindipity and you may want to conclude the lesson by reading the story. Unfortunately, the relationship, if any, between gravity and magnetism and electricity remains a mystery.

If compasses are available you can have students observe that electricity creates a magnet field. Wrap wire twice around the compass. Note the direction the compass is pointing. Connect the ends of the wire to the ends of the battery and suddenly the compass changes its direction. Why? When electricity flows through wire it creates a magnet field. Why? That answer awaits discovery.

MATERIALS
Per student or team
3 steel pins
1 magnet
1 compass (optional)

Ship Captains, Pins, and Magnetic Poles

Specific Concepts/Skills (Review)

1. Some things are magnets, while others are only attracted to magnets.
2. The magnetic properties of magnets are concentrated in two poles.
3. A piece of iron or steel can be magnetized by stroking a magnet with it.

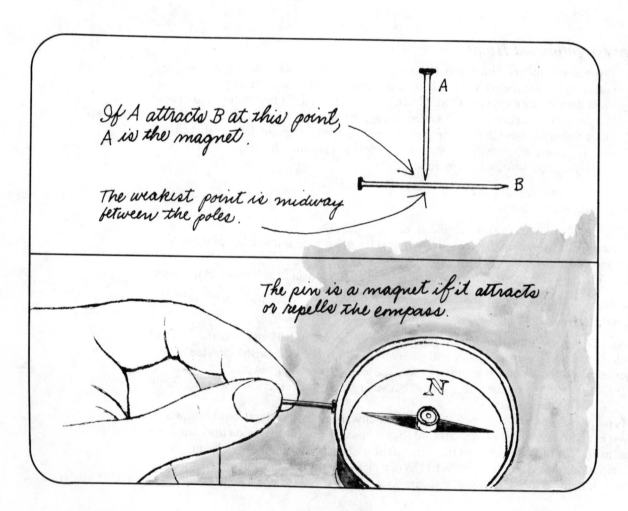

If A attracts B at this point, A is the magnet.

The weakest point is midway between the poles.

The pin is a magnet if it attracts or repells the compass.

Note to Teacher

This lesson reviews the concepts learned previously and tests the youngsters' ability to apply what they have learned in solving an interesting problem. You may wish to begin the lesson with an old and famous riddle that parallels what they will be doing. Present the riddle orally for mature youngsters, but write it on the board for others. It's not a bad idea to send the riddle home for parents to solve.

The Riddle

The captain of a sailing vessel approached an island surrounded with a coral reef. There was one opening in the reef and only the natives on the island knew where it was. There were two groups of natives, the Truthtellers and the Liars. The Captain knew that Truthtellers *always* told the truth and that Liars *always* lied.

As the captain's ship approached the reef, he saw three natives on shore. He called to them: "Are you Truthtellers?" One native replied, but the captain couldn't hear what he said. The captain called again, "What did he say?"

The second native said "He said he's a Liar, but I am a Truthteller."

The third native said, "No, he said he's a Truthteller, and I'm a Truthteller too."

Almost at once the captain knew which group each native belonged to. What group does each one belong to, and how did the captain know?

The captain was one smart fellow. It will take youngsters (and teachers) much longer to solve the riddle. Allow 5 or 10 minutes of discussion, then move on, permitting the youngsters to solve the riddle later on their own, or with a parent's help. You might offer a "fantastic" prize for the first one with a correct answer and *reasons* for the answer.

The Solution

A Liar or a Truthteller would always say he was a Truthteller. Therefore, the second native is a Liar. The third native told the truth, so he is a Truthteller. The word "too" assures the captain that the first native is also a Truthteller.

The Activity

1. Say: "I am going to give each of you two steel pins and a magnet. Don't let the pins touch the magnet. If you do, you probably won't be able to solve today's science problem. (Both pins may become magnetized.)
2. Pass out the materials.
3. Say: "Select either pin and make it into a magnet." (They should recall that stroking the pin in one direction across a magnet will turn the pin into a weak magnet.)

4. Say: "Test your pin to be sure it is a magnet." (It should lift or at least move the unmagnetized pin.) "Move your magnet away from the pins. You don't want the second pin magnetized."

5. Say: "One partner should mix the two pins up but don't let them rub against each other. One might magnetize the other."

6. Say: "Here is your science problem. Using only the two pins, can you find out which one is the magnet? If you say yes, you must explain how you decided. (This is a very difficult problem to solve using only the two pins. Many youngsters will lift one pin with the other and decide the pin doing the lifting is a magnet. However, if you ask them to try the reverse, it will also work. Almost no youngster can solve this problem and you should move on to step 7. But you may have a young Sir Isaac Newton, who reasons as follows: Magnets have poles. The strength of the magnet is concentrated there. As you move from one pole to the other, the strength will gradually diminish and then gradually increase. The point in the middle of the two poles will be magnetically balanced and not have magnetic properties. By applying this theory, the magnetic pin can easily be found. Select either pin, let's call it pin A. Touch its point to the end of the other pin (B) and lift. Whether A or B is the magnet, B will be lifted. Gradually move pin A toward the center of pin B. When the center is reached, either A will attract B or it won't. If A attracts B, A is the magnet. If A doesn't attract B, B is the magnet.)

7. Say: "I am now going to give each of you a third pin that is unmagnetized. Place the new pin with the other two and mix them. Now can you tell which one is the magnet?" (Two pins will not attract each other, and the third is a magnet.)

8. If compasses are available, say: "Can you use the compass to find out which pin is a magnet?" (Since the compass pointer is a magnet it has poles. When the end of the magnet pin is brought near the compass, the pointer will either point toward or away from the pin. If the pin is reversed, the pointer will also reverse.)

9. Discuss the answers and review.

Investigations at Home

1. Working out the riddle.
2. Allow each youngster to take one magnet home. Have them use the magnets to magnetize screwdrivers. Be sure to remind them to get permission first. There may be other things around the house to magnetize and with permission they can magnetize the others as well. Have them report to the class what they magnetized and you can then collect the magnets if you want to.

Evaluation

1. Approximately 6,000 students and teachers have tried this investigation and only one solved the problem in step 6. Did anyone in your class solve it? If so, please write me to let me know how their ideas developed.
2. Could almost every youngster find the magnetized pin using three pins or a compass?
3. Were some pins used for other kinds of investigations?
4. Were youngsters anxious to bring the magnet home? Did you let them? They only cost about eight cents.

Building the Shoestring Motor

Specific Concepts/Skills

1. An electrical motor is a device that changes electrical energy to mechanical energy by means of magnetism.
2. Building a motor takes patience and cooperation.

MATERIALS

Per motor

Approx. 1 m (3 ft) winding wire

2 paper clips (bent in advance for young children)

1 or more student magnets

1 Styrofoam cup

1 pair of scissors

1 test tube

1 1.5V battery

Small piece of sandpaper

Approx. 15cm (6 in.) of masking tape

1 rubberband

Per class

Extra winding wire

One pair of pliers (optional)

Extra batteries

Note to Teacher

Caution: This one investigation may lead to your being immortalized in the hearts and minds of your students. Being immortal carries some responsibility. Prepare yourself by building this motor in advance. Then gather enough materials for your class, select just the right hour, and introduce this lesson. Incidentally, don't be discouraged by people who know all about motors and assure you that the Shoestring Motor can't work. In theory it might not, but in fact it does!

Special thanks are due Lawrence Hall of Science, Berkeley, California, where I saw the prototype of this motor and to Aron Wong, a student of mine, who greatly improved my version of the Shoestring Motor.

Introducing the Activity

1. Perhaps the best way to begin this lesson is to display your motor. Review how you built it; then distribute the materials and hand out a copy of the instructions to each team. After the motors are running you might challenge the class with questions such as:
   a. In which direction does the coil spin, clockwise or counter clockwise? (Varies.)
   b. Can you find one or more ways to change its direction?
   c. Can you find one or more ways to increase the speed at which the coil spins? (Use a second battery connected in series; bring the magnets closer to the coil; use more magnets; use more wire when making the coil.)
   d. What do you think would happen if one magnet was placed on each side of the coil? Try it, but remember each magnet has an N and an S face.
   e. Use the motor to turn a wind dial or a color wheel.
   f. Find a way to change the motor and make it work better.

*Instructions for Building the Shoestring Motor

1. As you build this motor, relax.
2. Wind the wire around the test tube to form a coil. Leave about 7 centimeters (3 inches) of wire free at each end.
3. Remove the coil from the test tube.
4. Loop each free end of the wire twice around and through the coil (see top photo on page 182). This will prevent the coil from unwinding.
5. Completely remove the insulation from each free end of the wire. To do this, lay the wire flat on your desk. Rub it with sandpaper. Be sure to turn the wire so as to remove insulation from all surfaces. The wire will appear silver when the insulation is removed.
6. Lay the battery on the inverted Styrofoam cup. Tape the battery to the cup (see

*Thanks to Gene Perna and Mary Nalbandian of the Chicago Public Schools for 1985 modification.

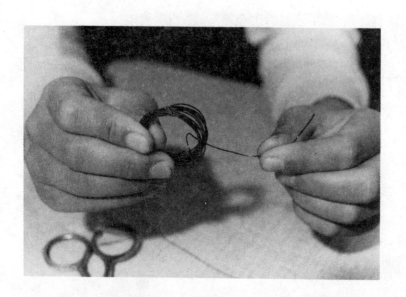

drawing, page 185).

7. Bend the paper clips as shown above. (Young children may need help with this step. If you have time, bend them in advance.)
8. Tape the clips tightly to the battery, as shown above. (Wrapping a rubber band around the clips and battery will help hold the clips tightly to the battery.)
9. Tape the magnets to the battery.
10. Lay the free ends of the coil onto the paper clips.
11. Spin the coil and it should continue to spin by itself. If it doesn't, spin it again. If the motor won't work, see the Trouble-Shooting Instructions.

Trouble-Shooting Instructions

1. If nothing happens, electricity may not be flowing through the coil.
   a. Try spinning the coil while the clips are squeezed against the battery.
   b. Check that the insulation is completely removed from the free ends of the coil.
   c. Try your coil on someone elses battery and clips. If it spins, check your battery, then repeat a and b.

2. If the coil rocks to and fro but won't spin
   a. Check balance of coil. Adjust free ends so they are wrapped around the center of the coil.
3. If there are screams and shouts
   a. Look around. Someone's motor is working just fine.
   b. Adjust the distance between the magnet and the coil by raising or lowering the paper clips. Be sure to retighten them against the battery.
   c. Try two magnets instead of one.

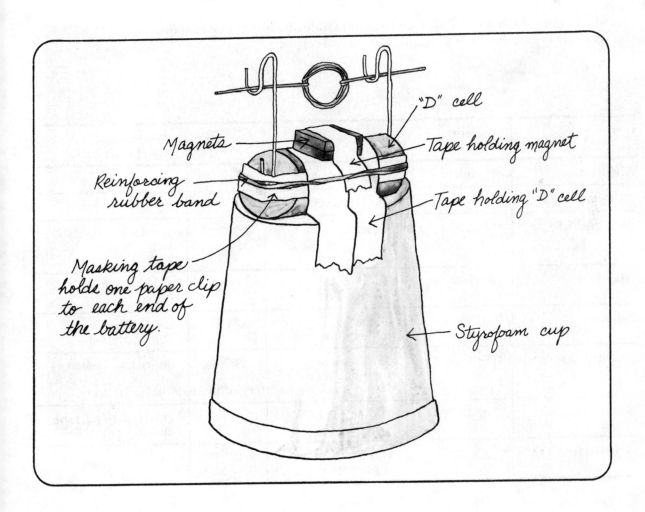

Magnets

"D" cell

Tape holding magnet

Reinforcing rubber band

Tape holding "D" cell

Masking tape holds one paper clip to each end of the battery.

Styrofoam cup

Appended list of materials

The materials included in this list are required for one or more investigations. It is suggested that they be secured in advance and stored in the science corner. Complete low cost *Science on a Shoestring* (SOS) kits as well as replacement materials are available from Learning Spectrum, 1390 Westridge, Portola Valley, CA 94025. See order form following page 28. Items marked with an asterisk* are used in more than one investigation and are essential.

The quantity of material is based on 30 students working in teams of two. This may be modified according to your needs and budget.

Items that are perishable or normally found in classrooms have been omitted from this list. They are included, however, with the **Materials** list on the first page of each lesson.

If you are ordering such items as clear plastic or Styrofoam cups in large quantities, you may want to purchase them wholesale from a local supplier.

Item	Quantity	Source	Approx. price	Comments
Alka-Seltzer	15 tablets	Supermarket	1.10	
*Aluminum foil	1 box	Supermarket	.65	
Bags, zip lock	1 box	Supermarket	1.20	Used for storage as well as for one investigation
*Balloons	35 9 inch	Toy or hobby store	5.00	Latex, small, easy to inflate
Balloons	2	Toy or hobby store	.25	Large, for teacher demonstration
*Battery, D cell	15–30	Hardware or supermarket Radio Shack has a battery-of-the-month club (free)	.35 each	30 are recommended
Battery, D cell	2 or 3	Hardware or supermarket	.35 each	Different brand from above
*Candle, five-inch	30	Hardware or supermarket	6.00	

Item	Quantity	Source	Approx. price	Comments
Candle, votive	30	Hardware or supermarket	6.00	May substitute for 5-inch candle
Clothespins	15 wide grasp	Hardware or supermarket	2.00	Used as test-tube holders and bulb holder
Coloring, food	1 small bottle each	Supermarket	.10 each	Red and green
*Container, with cover (plastic shoe box)	15	Supermarket or drug store	2.50 each	Necessary item: a clear top is preferred
Container, paper, waxed	30	Supermarket	3.00	5 oz and up; lip on the underside very useful
Corks	15	Hardware	.15 each	A Styrofoam cup may be substituted
Cotton	1 box	Supermarket	1.75	Paper towel or blotter paper may be substituted
*Cup, clear plastic, 5-oz	30	Supermarket	3.50	Any size from 5 oz to 9 oz will do; select cheapest available
Cup, paper	30	Supermarket	.01 each	Used in string phone investigation; substitute with any available cup
*Cup, Styrofoam	4	Supermarket	.02 each	
*Dropper, eye	15	Hardware, drugstore	.10 each	
*Lens, hand	30	Learning Spectrum	.30 each	5 power
*Lightbulb, flashlight #47 or #48	15–30	Hardware or electrical supply store	.45 each	Any flashlight bulb may be substituted

Item	Quantity	Source	Approx. price	Comments
Litmus paper	1 bottle red 1 bottle blue	Scientific supply house	1.00 each	
Magnet, dish	15	Toy, hobby, or electrical supply store (Radio Shack)	.18 each	Any other small magnet may be substituted
*Magnet (student), rectangular with hole	30	Toy, hobby, or electrical supply store. (Learning Spectrum)	.60 each	Very important item
*Matches, safety	4 boxes	Supermarket	.15	
*Matches, nonsafety	1 box	Supermarket	.45	
Mealworms	40	Pet shop	2.00 each	Order 100 and keep the rest
Needles, sewing	2	Supermarket	.10 each	About an inch long
Paper, black 8" x 10"	3 pieces	Supermarket		
Paper toweling	1 roll	Supermarket	.75	Used to substitute for cotton in seed growth investigation
*Paper clips	6 boxes	School	.30 each	Very useful; washers may be used as a substitute
Parachute paper	30 pieces	Supermarket	1.00 each	Apple wrap or tissue wrapping
Pins, brass-filled	2	Supermarket	.05	

Item	Quantity	Source	Approx. price	Comments
*Pins, steel-filled	90	Supermarket	1.25	Steel-filled pins are attracted to a magnet
Popcorn seeds	small bag	Supermarket	.60 can	Buy 2 kinds
Raisins	1 small box	Supermarket	.20	About 30 are needed
Rubber bands	60	Supermarket	.50	Various sizes
*Salt	1 box	Supermarket	.45	Very little needed
*Screw, iron or nail	15	Hardware	.03–.10 each	3 inches long; an iron nail may be substituted
Seeds	1 oz each kind	Supermarket or nursery	.25–.30 pkg	Alfalfa, mung
*Seeds, bird	½ lb	Supermarket or nursery	.50	Very useful; must not be sterilized
Seeds, lima bean	1 lb	Supermarket or nursery	.50	Any uncooked lima beans will do
Silly Putty	1	Toy or hobby store	1.00	
Steel wool	1 pad	Supermarket	.03	
Stick, skewers	15	School nurse	.50	Use a split tongue depressor or substitute
Straws	30	Supermarket	.15	Plastic or paper
*String	1 ball	Supermarket	.75	Any string
Sugar, brown	1 lb	Supermarket	.70	
*Tape, masking	1 or 2 rolls	Supermarket or hardware	.60 each	About ¾ inch wide; very useful

Item	Quantity	Source	Approx. price	Comments
*Test tube, large	15	Scientific supply house (Learning Spectrum)	2.50	You may want to have 30 on hand; approx. 18 × 100 mm borosilicate culture tubes
Vinegar	1 bottle	Supermarket	.60	Cheapest available
Water softener	1 oz	Supermarket		PhotoFlo recommended
Wax	4 lbs	Hobby shop	5.00	Cheapest dipping available in bulk
Waxed paper	1 box	Supermarket	.80	A substitute for waxed cup
Wire, insulated	100 ft	Hardware or Learning Spectrum	5.00	Narrow gage; for electromagnet and for SOS motor
Wire	100 ft	Hardware Learning Spectrum	5.00	For SOS motor
Wire, stripper	1	Hardware		Or borrow for one time use